Advances in Computer Vision and Pattern Recognition

Founding editor

Sameer Singh, Rail Vision, Castle Donington, UK

Series editor

Sing Bing Kang, Microsoft Research, Redmond, WA, USA

Advisory Board

Horst Bischof, Graz University of Technology, Austria
Richard Bowden, University of Surrey, Guildford, UK
Sven Dickinson, University of Toronto, ON, Canada
Jiaya Jia, The Chinese University of Hong Kong, Hong Kong
Kyoung Mu Lee, Seoul National University, South Korea
Yoichi Sato, The University of Tokyo, Japan
Bernt Schiele, Max Planck Institute for Computer Science, Saarbrücken, Germany
Stan Sclaroff, Boston University, MA, USA

More information about this series at http://www.springer.com/series/4205

Ajay Kumar

Contactless 3D Fingerprint Identification

 Springer

Ajay Kumar
The Hong Kong Polytechnic University
Kowloon, Hong Kong

ISSN 2191-6586   ISSN 2191-6594   (electronic)
Advances in Computer Vision and Pattern Recognition
ISBN 978-3-030-09807-0   ISBN 978-3-319-67681-4   (eBook)
https://doi.org/10.1007/978-3-319-67681-4

This Springer imprint is published by the registered company Springer Nature Switzerland AG
The registered company address is: Gewerbestrasse 11, 6330 Cham, Switzerland

Preface

Person identification using epidermal ridge impressions from fingers has been widely studied for over hundred years. It is widely employed in a range of forensic, e-business and e-governance applications around the world. Traditional acquisition of fingerprint images by rolling or pressing of fingers against hard surface like glass or polymer often results in degraded images due to skin deformations, slippages, smearing or residue of latent from previous impressions. As a result, full potential from the fingerprint biometric cannot be realized. Contactless 2D fingerprint systems have emerged to provide improved hygiene and ideal solutions to above intrinsic problems. Contactless 3D fingerprints can potentially provide significantly more accurate personal identification, as rich information is available from contactless 3D fingerprint images.

Contactless 3D fingerprints offer exciting opportunities to improve the user convenience, hygiene and the matching accuracy over the fingerprint biometric technologies available today. Introduction of videos, or addition of an additional temporal dimension, was a leap forward that revolutionized the usage of 2D images in the entertainment, e-governance and e-business. Similarly, the addition of one more dimension from 3D fingerprints, has potential to significantly alter the way this biometric is perceived and employed for the civilian and e-governance applications. Such advancements will not be limited in e-security or e-business, but also enable dramatic advancements in forensics where the latent or lifted fingerprint impressions are matched with suspects fingerprint images. For example, the 3D fingerprints from possible suspects can be employed to simulate latent fingerprint impressions on a variety of hard or soft real-life materials (door, paper, glass, gun, etc.) and under variety of pressure, occlusions and deformations, which is expected to enable more accurate match with the corresponding latent fingerprints that are lifted from the crime scene. The potential of contactless 3D fingerprints offers exciting opportunities but requires significant research and development efforts for its realization.

Availability of a book that is exclusively devoted to the techniques, comparisons and promises from the contactless 3D fingerprint identification is expected to help in advancing much needed further research in this area. Some of the contents in this book have appeared in some of our research publications and US patents. However, many of the important details, explanation and results that have been missed in the publications are included in this book. The contents in this book attempt to provide a systematic introduction to the 3D fingerprint identification, including most updated advancements in contactless 2D and 3D sensing technologies, and explanation of every important aspect towards the development of an effective 3D fingerprint identification system.

This book is organized into eight different chapters. Chapter 1 introduces current trends in the acquisition and identification of fingerprint images. This introductory chapter discusses the nature of fingerprint impressions and the sensing techniques, which includes completely contactless 2D fingerprint sensors. This chapter bridges the journey from rolled and inked fingerprint impressions, to the more advanced smartphone-based fingerprint sensors, in terms of their resolution and sensing area. It also provides details on publicly accessible implementations on fingerprint matchers and most updated list/details on publicly available fingerprint databases along with respective weblinks to enable easy accessibility.

Chapter 2 in this book presents a range of 3D fingerprint imaging techniques along with their comparative technical details. Image acquisition methods presented in this chapter have been grouped into four categories: optical, non-optical, geometric and photometric methods. Details on five different methods to acquire 3D fingerprint images using stereo vision, pattern lighting, optical coherence tomography, ultrasound imaging and photometric stereo, along with potential from *other methods*, appear in this chapter.

Chapter 3 in this book is devoted to in-depth details on a low-cost and effective method for the online 3D fingerprint image acquisition. Systematic details on such photometric stereo-based setup are detailed in this chapter, i.e. from hardware, calibration, preprocessing and specular reflection removal, to the choice of reconstruction methods. This chapter also shares our insights and results on the attempts made to consider non-Lambertian nature of finger surface. Resulting computational complexity for such online 3D fingerprint imaging system also appears in this chapter.

Chapter 4 provides details on more efficient 3D fingerprint imaging approach using coloured photometric stereo. This approach is introduced to address two key problems associated with practical 3D fingerprint imaging: involuntary finger motion and complexity for online applications. This approach revisits the method detailed in Chap. 3 and contactless 3D fingerprint images acquired using the setup introduced in this chapter are also publicly made available.

Contactless 3D fingerprint data often requires preprocessing operations to suppress the accompanying noise and to enhance or accentuate the ridge–valley features. Chapter 5 in this book details on such preprocessing operations on the cloud point 3D fingerprint data. This chapter also provides detailed explanation on specialized enhanced operations required for the contactless 2D fingerprint images that are employed for the reconstruction of 3D fingerprints.

Chapter 6 systematically introduces representation of minutiae in 3D space and provides details on recovering these features from cloud point data. Therefore, the techniques discussed in this chapter are generalized and quite independent of method used for the 3D imaging. With the help of many illustrations, most from real 3D fingerprint data, this chapter systematically details alignment and relative representation of 3D minutiae in order to generate numerical match score between two arbitrary 3D fingerprint minutiae templates. This chapter also details a minutiae selection algorithm and in-depth study on the variation of five-tuple relative 3D minutiae components with distance, which resulted in the introduction of a unified matching distance. Detailed experimental results presented in this chapter underline the effectiveness of our 3D minutiae template-based approach.

Contactless 3D fingerprints can be matched using a range of methods than those detailed in Chap. 6. Therefore, Chap. 7 details on such efficient 3D fingerprint matching methods, binary surface code-based approach and its variants, along with tetrahedron-based matching approach. Methods detailed in this chapter offer computationally efficient alternatives that can justify their usage in a range of e-business or civilian applications.

Chapter 8 in this book is devoted for the study on the uniqueness of 3D fingerprints. This chapter scientifically defines the individuality of 3D fingerprints and comparatively evaluates its improvement, over the 2D fingerprints, using practical 3D minutiae template matching criterion and imaging resolutions. The numbers illustrated in this chapter provide upper bound on the expected performance from the contactless 3D fingerprint systems.

I wish to thank many student and staff members in *The Hong Kong Polytechnic University*, who have directly or indirectly supported in the completion of this book. *Cyril Kwong* and *Chenhao Lin* deserve special thanks here as they have been instrumental in advancing many of the research outcome reported in this book. *Cyril* has worked with me for several months as research assistant, while *Chenhao* has been working towards his doctoral degree research.

Kowloon, Hong Kong   Ajay Kumar
June 2018

Contents

Chapter 1
Introduction to Trends in Fingerprint Identification

The finger skin patterns are widely considered as unique to humans, serve as the basis of forensic science and increasingly employed in large-scale national identification (ID) programmes for security and e-governance. The inked impressions of fingers on paper or the latent finger impressions on objects have been historically used to establish identity of individuals and commonly referred to as the fingerprint identification. However, modern imaging does not require such inked impressions. Therefore, the finger image identification is essentially same as fingerprint identification and is interchangeably used in this book.

The fingerprint ridges are essentially the combinations of *ridge units* which are combined under some random forces to form a continuous ridge flow patterns. The discontinuities in such ridge patterns are used to uniquely discriminate the fingerprints and commonly referred to as *minutiae*. These ridge discontinuities can appear as a pattern depicting tiny incomplete ridge spur, an abrupt bifurcation or abrupt termination of ridges. The spatial location and relationship between these known kinds of minutia types are unique for each person or even among his/her different fingers. The minutiae extracted from the fingerprint images can also be represented as connected graph whose nodes represent a known kind of minutia. The recovery and matching of such minutia patterns form the scientific basis of fingerprint identification. The process of *ridge units* formation has been scientifically linked [1] to the skin cells that are generated and periodically migrated towards epidermal surface. The formation of ridge patterns or the fingerprint starts before the birth and these patterns are already in the foetus during the fifth month of the pregnancy.

1.1 Contact-Based Fingerprint Identification

Traditional fingerprinting process requires the subjects to press or roll their fingers on a paper or on some hard surfaces generally made of a polymer or glass platen which forms front-end interface of a solid state sensor. The complexity of finger-

© Springer Nature Switzerland AG 2018

A. Kumar, *Contactless 3D Fingerprint Identification*, Advances in Computer Vision and Pattern Recognition, https://doi.org/10.1007/978-3-319-67681-4_1

printing process varies with the requirements of impression types and has been fairly standardized by the law enforcement agencies around the world.

(i) Rolled Fingerprint Impressions: The rolled fingerprint impressions can cover largest finger areas and acquires rich information which is known to achieve extremely accurate fingerprint matching. The rolled fingerprint impressions are individually acquired from each of the fingers by rolling them from nail-to-nail. The acquisition process is quite time-consuming and can also result in poor quality images due to uneven pressure from fingers. The rolled fingerprint impressions are widely used in the property documents and in prisons by the law enforcement departments.

(ii) Plain Impressions: This process of imaging does not require rolling of fingers and impressions are acquired pressing finger surface against a flat surface or a paper. Plain impressions from single fingers or thumb are least complex and widely used in e-governance (HK ID cards [2] and e-channel border crossings), e-commerce and smartphones.

(iii) Slap Impressions: These impressions can be simultaneously acquired for multiple fingers and this process is significantly faster particularly when ten prints from both hands are to be acquired. Tenprint plain impressions using slap fingerprinting is used in large-scale identification programmes, e.g. Aadhaar [3] in India or USVIST [4] programme in the USA, and consists of 4-4-2 method; acquiring slap impressions from the right hand, followed by a left hand and followed by those from two thumbs. The slap fingerprint images are often automatically segmented into plain fingerprint images from the individual fingers.

(iv) Latent Impressions: Latent fingerprints are leftover impressions of fingers on any object surface by the individual whose identity is yet to be established. The residue of such impressions is carefully lifted by the forensic experts using specialized techniques. These techniques typically use treatment, i.e. spraying the surface with chemicals, whose choice can vary depending on the nature of background surface, to enhance the impressions and acquiring photographs under ultraviolet illumination. The latent fingerprint impressions are generally of extremely low quality and with least clarity due to uncontrolled background. Therefore, the matching of latent fingerprint impressions is most challenging and also widely debated in courtroom arguments (Table 1.1).

Automated improvement of fingerprint image quality is employed in almost all the commercially available fingerprint sensors. In addition to the use of image enhancement algorithm to enhance clarity of ridge and valleys, several hardware-based solutions are also incorporated into many of these systems. Such enhancement in the quality of fingerprint images from dry skin is achieved with the use of a deformable membrane on glass platen (e.g. as in [5]) and/or with the use of a heated platen (e.g. as in [6]) to suppress or eliminate the undesirable influence of finger moisture. Many online fingerprint image acquisition software can also include detection of latent impressions left from previous scans, incomplete or partial images or finger slippage.

Table 1.1 Average number of minutiae recovered from typical fingerprint impressions

Impression type	Average number of minutiae	Sensor area
Rolled fingerprints	~80	~422 mm^2 [a]
Flat fingerprints	~20–30	211 mm^2 [b]
Latent fingerprints	~13–22[c]	–
Smartphone fingerprints	~5–8	50–100 mm^2

[a]Require about twice the area than for flat impressions and generally scanned from inked impressions
[b]At least 12.8 mm wide and 16.5 mm high (for 500 dpi as per ANSI/NIST/UIDAI Specifications [37], [8])
[3]Estimated from NIST SD27 [38] dataset which has a varying number of usable minutiae

Quality of a fingerprint image is often judged by the clarity of ridge patterns and is quantified by imaging resolution in pixels per inch (PPI). Several law enforcement departments (e.g. FBI [7]) and national ID programmes (e.g. Aadhaar [8]) have standardized the imaging requirements and require 500 dpi imaging sensors for their applications. Accordingly approved lists of commercially available sensors that meet respective criterion are publicly made available [5, 7] for the users. In order to acquire full plain impression of a fingerprint, the sensing area should be at least *one by one squared inches* and also has been standardized in these specifications. However, a variety of fingerprint sensors with lower resolutions 200–300 dpi are also employed in a range of e-business and stand-alone applications like for the office attendance or in the laptops. The choice of fingerprint sensor is generally a trade-off between the required/offered level of security, available sensing area, cost, speed and storage that are affordable for the respective application.

Interoperability among the fingerprint sensors is critical to the success of large-scale ubiquitous identification programmes. Such interoperability is facilitated with the help of standards and certification programmes. Aadhaar and FBI publicly provide a list of certified fingerprint scanners, in [9] and [7], respectively, that generates fingerprints with image quality that is in compliance with the expectations for the large-scale identification process. Commercially available contact-based fingerprint sensors employ a variety of sensing technologies and these can be grouped into following categories:

(a) Optical: The frontal side of imaging platen makes contact with the finger ridges during imaging while other side senses the light reflected from the ridges using CMOS or CCD sensor. The image formation in such sensors is based on the principle of total internal reflection where the lights reflected by the valleys appear brighter while the light randomly scattered by ridges creates darker impressions in the sensor images. Optical sensors are not susceptible to influences from electrostatic discharge in the surrounding but require relatively larger size. Optical scanning technology is the basis of majority of slap and/or stand-alone commercial fingerprint scanners available today.

(b) Silicon: Unlike optical sensors, the image formation in silicon fingerprint sensors relies on the capability of capacitive, thermal, electric or piezoelectric sensors in generating discriminative signals at spatial locations corresponding to the ridges and the valleys. Therefore, according to the nature of sensors, silicon fingerprint sensors are classified as capacitive, thermal, electric field or piezoelectric fingerprint sensors. Capacitive fingerprint sensors are insensitive to ambient illumination and more resistant to the contaminations. Capacitive fingerprint sensors, with silicon or polysilicon as the base material, also offer cost advantage and widely used in smartphones available today.

More recent research and development efforts result in development of *transparent* fingerprint image sensing from the entire display area of widely used smartphones which can offer higher user-friendliness by operating on *touch me anywhere* mode or provide simultaneous acquisition of *multiple* fingerprints with ease. Such fingerprint sensor can reside on the top of existing smartphone display but embedded under the cover glass. These fingerprint sensors also operate on capacitive difference between ridges and valleys introduced during human finger touch on smartphone displays. Figure 1.1 illustrates an example of such transparent fingerprint sensor, along with the sensed fingerprint images, from [10]. Such fingerprint sensing can achieve more than 500 dpi of resolution, enabling the acquisition of pores, and offer an attractive alternative for a range of e-business and law enforcement applications.

(c) Ultrasound: Ultrasound fingerprint sensors use response from acoustic signals from finger ridges to reconstruct the fingerprint images. High bulk and cost are the key reasons that such ultrasound fingerprint sensors are hardly used in real applications. New ultrasonic fingerprint sensors, e.g. sense ID ultrasonic fingerprint sensor [11]. Qualcomm Fingerprint Sensor and Lamberti et al. [12, 13] have been recently been introduced to address these limitations for mobile applications.

(d) Multispectral: Multispectral fingerprint sensors simultaneously image the fingers at different wavelengths to generate composite fingerprint image. The multispectral imaging attempts to simultaneously recover subsurface and surface characteristics. The combination of such characteristics can improve image quality for dry and moist fingers. Multispectral imaging can generate superior quality images but is more complex and costly.

(e) Swipe Sensors: These sensors use small rectangular area and the users are required to sweep his/her finger on this area. The sensor reconstructs fingerprint image from the image slices that are acquired during the finger movement. These sensors require smallest area and do not suffer from the problem of leftover latent impressions. Swipe sensors require complex user interaction and additional image reconstruction process which can influence the accuracy and applicability of these sensors.

Fig. 1.1 **a** Transparent fingerprint sensor from [10] and **b** sample fingerprint images from this sensor

(a)

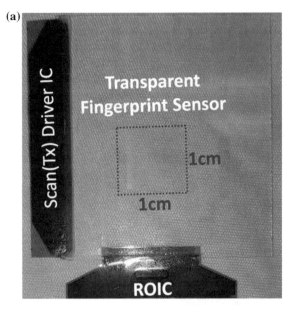

(b)

1.1.1 Matching Fingerprint Images

Fingerprint matching has been widely debated in courtroom arguments, and matching of two fingerprints is based on premise or uniqueness of matching usable minutiae patterns in their relative spatial positions. Establishing identity of suspects using latent fingerprint can be very difficult especially when only a portion of the finger is present in such latent impressions where the clarity of ridge patterns is blurred or occluded. Such latent impressions are not uncommon and present complex shape and configuration of ridge patterns where even certified fingerprint examiners can have a different opinion [14] on existence, location and type of minutiae. This is the key reason that automated decisions from the matching of latent fingerprints, with rolled or flat fingerprint for forensic examination, is not used in practice particularly for low quality of latent fingerprints. Automated systems like AFIS [15] while operating with latent fingerprints generate list of top-k, where k is generally smaller than 50, matched candidates for human examiners to compare and make decisions. This is also the main reason that the accuracy of latent fingerprint matching algorithms is generally reported in identification rate, rather than using receiver operating characteristics from verification experiments, due to nature of applications/deployments.

The flat or rolled fingerprint matching algorithms are largely based on uniqueness in spatial localization of various minutiae types in the fingerprints that are matched. However, recovery and reproducibility of every minutia in individual's fingers from the fingerprints acquired using modern or real imaging devices cannot be guaranteed. Therefore, most successful approaches for fingerprint matching accommodate adverse (but frequent) impact of missing minutiae and quality of recovered minutiae in their algorithms. There are a range of publicly available implementations for the quantifying the fingerprint image quality. Among these, the open source NFIQ 2.0 (NIST Finger Image Quality) [16] is most recent, popular, and quantifies the fingerprint image quality in 0–100 range according to [17]. NFIQ 2.0 is however specifically developed to quantify image quality for plain fingerprint impressions acquired from optical sensors or scanned inked impressions and should *not* be used for other sensing technologies like contactless fingerprints, etc. There are also publicly available implementations of fingerprint image matching algorithms that are based on most reliable minutiae features. Among these, those provided from NIST, i.e. NBIS [18] which includes NFSEG for the fingerprint segmentation, MINDCT to locate and detect minutiae features, and BOZORTH3 to generate match scores from MINDCT feature templates, include source codes and are more popular. A large-scale evaluation of fingerprint matching algorithms from different vendors was conducted by NIST, which involved comparisons between 10,000 matching subjects and 20,000 nonmatching subjects fingerprint images in a database of about 10 million subjects. The detailed report on this assessment was publicly released in 2015 [19] and provided a comparative summary of performance using identification accuracy and also the matching speed. A large-scale evaluation of fingerprint matching performance also appears in [20] which involved 84 million different subjects in the gallery. Large-scale evaluation results in these two reports: one-to-one fingerprint matching in [19] while one-to-many fingerprint matching in [20], using database of millions of subjects which suggests that a lot of further work is required to improve the capabilities of fingerprint matchers before these can be considered for stand-alone high-security applications.

Development of accurate, fast and fully automated fingerprint matching algorithms has attracted significant attention in academia, government and industry. Several fingerprint databases have been made publicly available to promote such development and advance research in this area. Table 1.2 summarizes such publicly available databases, along with their references that can be used to access or download them for research and development purpose. Reference [51] provides details on multisensory optical and latent fingerprint databases in [39]. Another more recent NIST fingerprint database from nail-to-nail fingerprint challenge, with a series of rolled and plain fingerprint impressions, is available from Ref. [52].

Table 1.2 Summary of contact-based fingerprint image databases in public domain

Impression type	Name of database	No. of subjects	No. of images	Image resolution	Reference to access
	NIST Special Database 27-latent	258	2580	800 × 768	[38]
Latent	IIIT-D Multi-sensor Optical and Latent Fingerprint (MOLF)-DB4	100	4000	Variable	[39]
	IIIT-D Multi-sensor Optical and Latent Fingerprint (MOLF)-DB5	100	1600	1924 × 1232	[39]
	NIST Special Database 10-plain	N/A	5520	832 × 768	[40]
	NIST Special Database 14-plain	13,500	27,000	N/A	[41]
	NIST Special Database 27-rolled	258	2580	800 × 768	[42]
Rolled	NIST Special Database 29-rolled	216	2160	N/A	[43]
	CASIA-Fingerprint V5	500	20,000	328 × 356	[44]
	CASIA Fingerprint Image Database for Testing Version 1.0	500	20,000	328 × 356	[44]
	NIST Special Database 300	888	8871	N/A	[45]
	IIIT-D Multi-sensor Optical and Latent Fingerprint (MOLF)-DB1	100	4000	352 × 544	[39]
	IIIT-D Multi-sensor Optical and Latent Fingerprint (MOLF)-DB2	100	4000	258 × 336	[39]
	IIIT-D Multi-sensor Optical and Latent Fingerprint (MOLF)-DB3	100	1200	1600 × 1500	[39]
	IIIT-D Multi-sensor Optical and Latent Fingerprint (MOLF)-DB3_A	100	4000	variable	[39]
	FVC2000-DB1	N/A	880	300 × 300	[46]
	FVC2000-DB2	N/A	880	256 × 364	[46]
	FVC2000-DB3	N/A	880	448 × 478	[46]

(continued)

Table 1.2 (continued)

Impression type	Name of database	No. of subjects	No. of images	Image resolution	Reference to access
Flat	FVC2000-DB4	N/A	880	240 × 320	[46]
	FVC2002-DB1	N/A	880	388 × 374	[47]
	FVC2002-DB2	N/A	880	296 × 560	[47]
	FVC2002-DB3	N/A	880	300 × 300	[47]
	FVC2002-DB4	N/A	880	288 × 384	[47]
	FVC2004-DB1	N/A	880	640 × 480	[48]
	FVC2004-DB2	N/A	880	328 × 364	[49]
	FVC2004-DB4	N/A	880	288 × 384	[49]
	FVC2006-DB1	N/A	1800	96 × 96	[49]
	FVC2006-DB2	N/A	1800	400 × 560	[49]
	FVC2006-DB4	N/A	1800	288 × 384	[49]
	NIST Special Database 29-plain	216	2160	N/A	[43]
	NIST Special Database 34—Plain and Rolled fingerprints	10	N/A	N/A	[50]
	NIST Special Database 300	888	8787	N/A	[45]

1.2 Contactless 2D Fingerprint Identification

Availability of low-cost contact-based fingerprint sensors has significantly contributed to the popularity of fingerprint identification for a variety of applications in e-business and e-governance. However, such conventional contact-based fingerprint sensing suffers from several limitations and the available performance from the state-of-art fingerprint matching algorithms [19] is much below the theoretical estimation and the expectations. Traditional contact-based acquisition requires the users to press or roll their fingers against the hard surface (glass, silicon and polymer) or paper. This process often results in partial or degraded quality fingerprint images which can be attributed to the improper finger placement, elastic skin deformation, slippages, smearing or to the sensor noise. Live fingerprint scans employed for commercial and law enforcement applications often have to cope up with the residue of dirt, moisture and sweat left from the previous fingerprint scans, e.g. manual cleaning of sensor surface. The non-contact, i.e. contactless or touchless fingerprint sensors can avoid the direct contact between imaging sensor and elastic finger surface. Such contactless fingerprint imaging approach can avoid the deformation of finger ridge patterns and acquire more acquire ground truth information. Therefore in addition to significantly improve hygiene, more accurate matching performance is expected from such contactless to contactless fingerprint matching. Figure 1.2 illustrates two arrangements of contactless finger imaging setup. The arrangement in Fig. 1.2a ensures that there is no contact between finger and sensor/camera and therefore fulfils the requirements of contactless fingerprint sensor just as in Fig. 1.2b. As compared with the sensor in Fig. 1.2a, the lack of support for sensor in Fig. 1.2b requires that the arrangement of imaging lenses to be chosen such that it can offer larger field of view and depth of focus so that acceptable quality images can be acquired with a relatively greater degree of freedom for the finger. This essentially means requirement of more complex arrangement of lenses for Fig. 1.2b, which increases the cost of system and also its vulnerability to optical distortions. Therefore, restricting the degree of freedom with arrangement in Fig. 1.2a can ensure that finger remains relatively focused with simplified and low-cost arrangement of lenses for contactless fingerprint imaging.

(a)   **(b)**

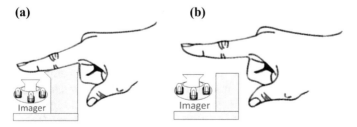

Fig. 1.2 Contactless fingerprint image acquisition setups: The setup in a uses *support* to limit the range of finger movement which can reduce the cost of sensor or camera optics. The setup in **b** requires larger depth of focus and view, which adds to the cost of contactless fingerprint sensing

Commercially available contactless fingerprint sensors can acquire single finger [21] contactless fingerprint images or simultaneously acquire multiple fingers [22, 23] contactless images. The technologies available for the contactless 2D fingerprint imaging can be broadly identified under two categories.

(a) Reflection Type: These types of contactless fingerprint sensors acquire images from the light reflected by LED illuminators that are placed in the front or on the same side as the imaging camera. Majority of the incident illumination is reflected by the finger ridges and ridges appear as bright pattern in the image. The illumination source is generally placed closer to the fingers so that the incident light is parallel to the optical path of camera. Incorrect placement of illumination, either for distance or angle, can generate shadows [24] that are projected from neighbouring 3D ridges. These unwanted shadows can alter the accuracy of minutiae localization due to the apparent shift in the profile of inter-mediate ridge–valley structure. The frequency of illuminator should be selected to ensure that there is minimum absorption by skin surface, lowest haemoglobin absorption, while the sensitivity of CMOS or CCD sensor is maximum. This compromise [25] often results in selection of *blue* light for the reflection type contactless fingerprint sensors. Figure 1.2 illustrates examples of commercial contactless fingerprint sensor that uses reflection type imaging.

(b) Transmission Type: The transmission type contactless fingerprint sensors acquire images from the frontal side of fingers while the illumination source is placed behind or towards fingernail side of the finger. Therefore, the frequency of illumination source is selected such that the transmission of illumination through the skin tissues is high. The penetrating illumination is imaged by the camera sensor placed on the other side of the finger. The motivation for the transmission type arrangement of the illuminators is to acquire interior or sub-surface details of the finger and to reduce the adverse impact of deterioration on skin or ridge–valley conditions. Reference [26] provides details on the devel-opment of transmission type contactless fingerprint sensor that uses 600 nm illuminator to acquire ridge–valley patterns from subsurface skin imaging.

Another transmission type contactless fingerprint imaging, with similar arrangement as in Fig. 1.2a, is based on optical coherence tomography (OCT) . Such OCT-based images can reveal subsurface fingerprint patterns up to the depth of 2–3 mm below the finger skin. The epidermis is the outermost layer of finger skin and is widely imaged for fingerprint impressions using the reflection type imaging. The finger skin layer *beneath* the epidermis is dermis. The upper side of dermis layer represents the papillary layer, which presents the *same* ridge patterns as observed from exter-nal fingerprints, and is referred to as the *internal* fingerprints in the literature. The OCT-based fingerprint imaging can acquire such *internal* fingerprints, which are better preserved and generally immune to finger surface dirt or damages to external ridge [27]. In addition, OCT-based fingerprint acquisition can also enable liveness detection as it can acquire capillary blood flow details [28] and therefore can offer an attractive alternative for high-security applications. However, current sensor cost for the OCT-based fingerprint acquisition is very high (in the order of tens of thousands of US$) and limits its use in popular e-business or security applications. There has

(a) **(b)** **(c)**

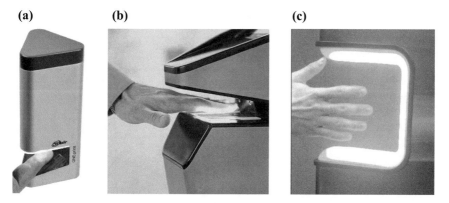

Fig. 1.3 Sample images of commercial contactless 2D fingerprint sensors: **a** touchless fingerprint reader *IDOne* [21], **b** on the fly from [23] and **c** finger on go from [22]

been some recent success in reducing the cost of such sensing and acquiring large area subsurface 2D fingerprints using full-field OCT. Reference [29] details acquisition of 1.7 cm × 1.7 cm fingerprint area with a resolution of 2116 dpi s in 0.21 s. Such efforts can be considered as remarkable advancement and such full-field OCT can be a cost-effective alternative as it uses an inexpensive camera and a light source to acquire OCT images.

The choice of illumination, its content and positioning, expected distance between the sensor and fingers, along with the positioning and optics of camera should be judiciously chosen to optimize the grey-level difference, i.e. image contrast, between the ridges and valleys in the acquired contactless fingerprint images. Despite such efforts, contactless fingerprint imaging generally results in lower contrast between ridge–valley structures. Therefore, additional contrast enhancement is required, before conventional fingerprint enhancement algorithms. This additional contrast enhancement typically employs homomorphic filtering and is detailed in Chap. 3. Another limitation with contactless fingerprint imaging is associated with the reduction in *size* of effective fingerprint area for identification. The perspective distortion in the camera, for the portions of curved finger skin that are far away from the centre, decreases the ridge–valley separation. Therefore, advanced algorithms are required to enhance and/or correct frequency varying ridge–valley regions. Several research efforts in academia [30] and industry have resulted in the simultaneous acquisition of five contactless 2D fingerprints, e.g. images in Fig. 1.3. Contactless fingerprint images require specialized algorithms for the preprocessing, which consists of image enhancement and image normalization, before incorporating the conventional approach [18] to generate minutiae-based fingerprint templates for the matching.

1.3 Contactless 3D Fingerprint Identification

Contactless 2D fingerprint acquisition offers deformation free acquisition of ground truth information and is expected to deliver superior matching accuracy fingerprint images acquired using contact-based sensors, which often suffer from sensor surface

Table 1.3 Comparison between touch-based, contactless 2D and contactless 3D fingerprint identification

	Touch-based 2D fingerprint	Contactless 2D fingerprint	Contactless 3D fingerprint
Recognition accuracy	High	High	Very High
Security hazards with sensor usage	High	Very Low	Very Low
Skin deformation	High	NIL	NIL
Sensor surface smear/noise	High	Very Low	Very Low
Identification of spoof and alterations	Low	Medium	High
Sensor cost	Low	High	Very High
Bulk/size	Compact	Medium/large	Bulky

noise, dirt, deformations, slippages or leftover latent from previous users. Contactless 3D fingerprint identification, in addition to providing key benefits associated with contactless 2D fingerprint images, can also incorporate additional information from 3D fingerprint surface. The use of such additional 3D geometrical information, like height, depth or curvature information, is expected to further improve the identification accuracy using 3D fingerprints. Traditional fingerprint identification using contact-based fingerprint sensors uses ridge information to generate minutiae templates while details on the valley are considered as background and therefore discarded. However, contactless fingerprint imaging can also preserve such valley details and can also be used to further enhance identification accuracy for contactless 2D and contactless 3D fingerprint matching. Table 1.3 presents a comparative summary of the expected benefits and indicative performances for contact-based, contactless 2D and contactless 3D fingerprint technologies. Replacement of contact-based fingerprint identification by contactless fingerprint identification can offer higher throughput, user convenience and hygiene in a range of e-governance applications. However, such replacement can also introduce challenges relating to accessibility, ergonomics, anthropometrics and user acceptability. A recent report from NIST in [31] presents a comparative study on a range of human factors *between* usage of contact-based and contactless 2D fingerprint sensors shown in Fig. 1.2b–c. These findings are quite encouraging and could lead to gradual replacement or deployment of contactless fingerprint sensors for e-governance applications in coming years.

Personal identification using contactless 3D fingerprint image has attracted the attention of researchers in the last one decade, and several efforts have been published [32–35] in the literature. However, there can be some confusion between fingerprint identification capabilities introduced from the 2D images acquired from *different* 3D views of a given finger and 3D fingerprint systems that acquire or use the depth information in contactless 3D fingerprint images, as both of these identification approaches have been referred to as 3D fingerprint identification in the literature. Use of multiple 2D fingerprint images from the same finger but using different 3D views appear in [32, 35]. Such an approach generates rolled equivalent of fingerprints along its 3D shape and uses multiple but fixed cameras to generate such rolled fingerprint image in a contactless manner. The rolled fingerprint impressions [36] can cover large 3D surface area and therefore generate fingerprint templates with a relatively large number of minutiae (Table 1.1). Therefore, similar 2D fingerprint templates, using multiple 3D views/cameras, can also lead to more accurate identification. However, true 3D fingerprint images are expected to include depth information corresponding to the fingerprint ridges, or at least the shape of 3D finger surface. Such contactless 3D fingerprint images are expected to offer significantly higher (Table 1.3) matching accuracy as these technologies are expected to discriminate the most reliable minutiae features in 3D spaces. There have been some interesting efforts to develop 3D fingerprint imaging capabilities and will be discussed in the next chapter.

References

1. Chen Y (2009) Extended feature set and touchless imaging for fingerprint matching, Ph.D. Thesis, Michigan State University
2. The Smart Identity Card (2018) Immigration Department, Hong Kong, https://www.immd.gov. hk/eng/services/hkid/smartid.html
3. Aadhaar Authentication, https://www.uidai.gov.in/authentication/authentication-overview/authentication-en.html 2018
4. US VISIT—Entry/Exit System (2016) http://www.immihelp.com/visas/usvisit.html
5. Secure Outcomes, LS 110 Fingerprint Sensor, http://www.secureoutcomes.net
6. Cross Match Technologies Inc., Guardian R2 Fingerprint Sensor, http://www.crossmatch.com
7. FBIs Certified Product List (CPL), https://www.fbibiospecs.org/IAFIS/Default.aspx
8. UIDAI Biometric Device Specifications, BDCS(A)-03–08, May 2012, http://www.stqc.gov. in/sites/upload_files/stqc/files/UIDAI-Biometric-Device-Specifications-Authentication-14-05-2012_0.pdf
9. List of UIDAI Certified Biometric Authentication Devices, Feb. 2018, https://uidai.gov.in/ images/resource/List_of_UIDAI_Certified_Biometric_Devices_13072017.pdf
10. Seo W, Pi J.-E, Cho S. H, Kang S.-Y, Ahn S.-D, Hwang C.-S, Jeon H.-S, Kim J.-U, Lee M. (2018) Transparent fingerprint system for large flat panel display. Sensors 293. https://doi.org/ 10.3390/s18010293
11. Tang H, Lu Y (2016) 3-D ultrasonic fingerprint sensor-on-a-chip. IEEE J Solid-State Circuits 51:2522–2533
12. Qualcomm Fingerprint Sensor, https://www.qualcomm.com/solutions/mobile-computing/ features/security/fingerprint-sensors. 2017
13. Lamberti N, Caliano G, Iula A, Savoia AS (2011) A high frequency cmut probe for ultrasound imaging of fingerprints. Sens Actuators, A 172(2):561–569
14. Ulery BT, Hicklin RA, Roberts MA, Buscaglia J (2016) Interexaminer variation of minutiae markup on latent fingerprints. Forensic Sci Int 246:89–99
15. FBI IAFIS CJIS Division, http://www.fbi.gov/about-us/cjis/fingerprints_biometrics/iafis/iafis_ latent_hit_of_the_year. 2017
16. NFIQ 2.0, NIST Fingerprint Image Quality, https://www.nist.gov/services-resources/software/ development-nfiq-20. April 2018
17. ISO/IEC. IS 29794-1:2016, information technology biometric image sample quality Part 1: Framework. ISO Standard Jan. 2016
18. NIST biometric image software (NBIS), Release 5.0 (2015) https://www.nist.gov/services-resources/software/nist-biometric-image-software-nbis

19. Watson C, Fiumara G, Tabassi E, Cheng S.-L. Flanagan P. Salamon W (2014) Fingerprint vendor technology evaluation, evaluation of fingerprint matching algorithms. NIST Interagency Report 8034. http://dx.doi.org/10.6028/NIST.IR.8034
20. Role of Biometric Technology in Aadhaar Enrollment, UIDAI, India, Jan 2012
21. ANDI ONE—Touchless Fingerprint Reader, http://www.andiotg.com/andi-oneprint
22. ANDI GO Zero Contact Fingerprint Identification System, Advanced Optical Systems Inc., http://www.andiotg.com/
23. https://www.idemia.com/sites/corporate/files/morphowave-tower-brochure-012018.pdf. May 2018
24. Parziale G, Chen Y (2009) Advanced technologies for touchless fingerprint identification. In: Tistarelli M, et al (eds), Handbook of remote biometrics. Springer–Verlag, London
25. Song Y, Lee C, Kim J (2004) A new scheme for touchless fingerprint recognition system. In: Proceedings of international symposium *on* intelligent signal processing and communication systems, pp 524–527
26. Sano E, Maeda T, Nikamura T, Shikai M, Sakata K, Matsushita M (2006) Fingerprint authentication device based on optical characteristics inside finger. In: Proceedings of CVPR 2006 Biometrics Workshop, pp 27–32
27. Darlow LN, Conan J, Singh A (2016) Performance analysis of a hybrid fingerprint extracted from optical coherence tomography fingertip scans. In: Proceedings of ICB 2016. Sweden
28. Liu G, Chen Z (2013) Capturing the vital vascular fingerprint with optical coherence tomography. Appl Opt 52(22):5473–5477
29. Anksorius E, Boccara AC (2017) Fast subsurface fingerprint imaging with full-field optical coherence tomography system equipped with a silicon camera. arXiv:1705.06272v2, https://arxiv.org/abs/1705.06272
30. Noh D, Choi H, Kim J, (2011) Touchless sensor capturing five fingerprint images by one rotating camera. Optical Eng 50:113202
31. Furman S, Stanton B, Theofanos M, Libert JM, Grantham J (2017) Contactless fingerprint device usability test. NIST IR Mar. https://doi.org/10.6028/NIST.IR.8171
32. Parziale G, Diaz-Santana E, Hauke R (2003) The surround imager: a multi-camera touchless device to acquire 3d rolled-equivalent fingerprints. In: Proceedings of ICB 2006, vol. 3832. LNCS
33. Wang Y, Hao Q, Fatehpuria A, Lau DL, Hassebrook LG (2009) Data acquisition and quality analysis of 3-Dimensional finger-prints. Proc IEEE Conf Biometrics, Identity and Security, Tampa, Florida, Sep, pp 22–24
34. Wang Y, Lau DL, Hassebrook LG (2010) Fit-sphere unwrapping and performance analysis of 3D fingerprints. Appl Opt 49(4):592–600
35. Touchless Biometrics Systems, Switzerland, https://www.tbs-biometrics.com/en/3d-enrollment/. 2018
36. Recording Legible Fingerprints (2016) https://www.fbi.gov/about-us/cjis/fingerprints_biometrics/recording-legible-fingerprints
37. American National Standards for Information Systems—Data Format for the Interchange of Fingerprint, Facial, & Other Biometric Information—Part 2 (XML Version); ANSI/NIST-ITL 2-2008, NIST Special Publication #500-275; National Institute of Standards and Technology, U.S. Government Printing Office, Washington, DC, 2008. Available online at http://www.nist.gov/customcf/get_pdf.cfm?pub_id=890062
38. Fingerprint Minutiae from Latent and Matching Tenprint Images, NIST Special Database 27, Gaithersburg, MD, USA, http://www.nist.gov/srd/nistsd27.cfm
39. IIIT-D Multi-sensor Optical and Latent Fingerprint (MOLF)-DB4, http://iab-rubric.org/resources/molf.html
40. NIST Special Database 10, NIST Supplemental Fingerprint Card Data (SFCD), https://www.nist.gov/publications/nist-special-database-10-nist-supplemental-fingerprint-card-data-sfcd-special-database. Feb. 2017
41. NIST Special Database 14-plain, http://www.nist.gov/srd/nistsd14.cfm
42. NIST Special Database 27-rolled, http://www.nist.gov/srd/nistsd27.cfm

43. NIST Special Database 29-rolled, http://www.nist.gov/srd/nistsd29.cfm
44. CASIA-Fingerprint V5 Database, http://biometrics.idealtest.org/dbDetailForUser.do?id=7
45. Fiumara G, Flanagan PA, Grantham J, Bandini B, Ko K, Libert J (2018) NIST Special Database 300, Uncompressed Plain and Rolled Images from Fingerprint Cards. NIST Tech. Notes 1993, June 2018, https://www.nist.gov/itl/iad/image-group/special-database-300
46. FVC 2000, http://bias.csr.unibo.it/fvc2000/
47. FVC 2002, http://bias.csr.unibo.it/fvc2002/
48. FVC 2004, http://bias.csr.unibo.it/fvc2004/databases.asp
49. FVC 2006, http://bias.csr.unibo.it/fvc2006/databases.asp
50. Flanagan PA (2018) Special Database 34—plain and rolled fingerprints from live and rescanned images, NIST Interagency/Internal Report (NISTIR)—8024, https://www.nist.gov/publications/special-database-34–150-plain-and-rolled-fingerprints-live-and-rescanned-images. April 2018
51. Sankaran A, Vatsa M, Singh R (2015) Multisensor optical and latent fingerprint database. IEEE Access
52. Fiumara G, Flanagan P, Schwarz M, Tabassi E, Boehnen C (2018) National institute of standards and technology special database 301: nail to nail fingerprint challenge dry run, NIST Technical Report No. 2002, July 2018

Chapter 2
3D Fingerprint Image Acquisition Methods

Acquisition of live and contactless 3D fingerprint images is an integral part of any 3D fingerprint identification system. Several 3D sensing technologies can be incorporated to acquire live 3D fingerprint images. This chapter introduces such 3D sensing technologies and provides a comparative assessment on the suitability of these technologies with respect to the factors like imaging principle, acquisition speed, complexity and precision.

The term *3D fingerprint imaging* here is referred to the process of acquiring 3D information from presented fingers and this information can include 3D shape of fingers, height/depth of 3D fingerprint ridge/valley or 3D shape of sweat pores. Acquisition of 3D fingerprint data should at least ensure that the depth information pertaining to fingerprint ridges are acquired from the proximal phalanx of the live finger presented by the user. In this context, it is useful to note that there is a range of systems [1] that can generate 3D representation from 2D grey levels of 2D fingerprint images and can be used to assist comparison between 2D fingerprint images for the forensic analysis. In addition, there are also multi-view techniques [2] that acquire multiple 2D fingerprint images from different 3D views of a presented finger. Such techniques lack precision in acquiring ridge depth information and are, therefore, not attractive to realize full potential from 3D fingerprint identification. The introduction in this chapter does not attempt to summarize the whole branch of research achievements in the acquisition of 3D fingerprint images but rather serves as an introduction to the topic and provides a comparative analysis of methods available in this area. Figure 2.1 presents a taxonomy of candidate 3D fingerprint acquisition methods that deserve consideration while developing a 3D fingerprint identification system. These methods can be largely classified into active or passive methods. Active methods require the projection of patterns, illumination or energy to record the 3D fingerprint details by evaluating the *response* (absence, presence and distortion) from the 3D surface to the input. The passive methods on the other hand record 3D information from the optical appearance of 3D fingerprint under focussed illumination. The active 3D fingerprint acquisition methods also include non-optical techniques like 3D ultrasound imaging.

© Springer Nature Switzerland AG 2018
A. Kumar, *Contactless 3D Fingerprint Identification*, Advances in Computer Vision and Pattern Recognition, https://doi.org/10.1007/978-3-319-67681-4_2

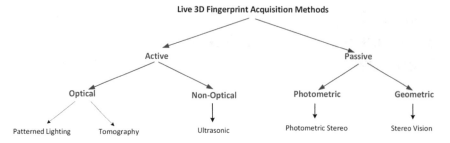

Fig. 2.1 Overview of existing techniques to scan 3D fingerprints images from a live user

2.1   Stereo Vision

Stereo vision based methods use at least two cameras to simulate human binocular vision and compute the depth information during live 3D fingerprint scans. The distance between the two cameras for such an approach is generally the distance between the human eyes, or intra-ocular distance of about 6.35 cm and greater distances can be employed for higher 3D details. Stereo vision based systems compute the depth information using the principle of triangulation. Active triangulation approach to scan 3D fingerprints can use a digital camera that records the response from a known laser signal, after its projection to 3D finger surface. The location of laser beam source, the camera and the contact-point of the laser beam to the 3D fingerprint surface forms a *triangle* and can be used to compute the depth information from such geometrical relationship. An example of such triangulation-based range sensor is Vivid 910 [3] which can offer accuracy up to one-tenth of a millimetre and has been used for biometrics research, e.g. to acquire palmprint data in [4]. Active triangulation-based range scanning methods available today generally illustrate high-frequency noise and its accuracy is limited in fixed multiples of sampling frequency. Motion of the fingers during the image acquisition process also limits the use of such range imaging methods for the practical 3D fingerprint identification.

In passive triangulation-based stereo vision methods, the laser (active) source is replaced by a second camera, which forms *passive* component in the triangulation. Such triangulation stereo, or shape from silhouette, based methods generally employ more than two cameras to enable nail-to-nail acquisition of 3D fingerprint shape details as shown in Fig. 2.2. The stereo vision theory explains [5] that the spatial location of an unknown object point in 3D space can be computed from the 2D images acquired at a different plane, at the same time. Such an approach also requires that the spatial locations of the camera and their correspondingly matching spatial points, referred to as *correspondence points*, are available. In case of multiple camera-based 3D fingerprint acquisition, the stereo image pairs are acquired from the adjacent cameras. The surround-imager in [6] employed five cameras while imaging system in [7] and [35] uses two cameras. In order to acquire nail-to-nail representation [8] of fingerprints, the arrangement in [6] employs *simultaneously* acquired five images from different 3D views. Such an arrangement uses silhouettes from each of the five-segmented images to generate a cylindrical model for the presented finger.

Passive triangulation-based methods to scan live 3D fingerprints using multiple cameras are simple to implement, fast and can acquire nail-to-nail *representation* of 3D fingerprints. Such shape from silhouette approaches can, however, only provide the shape of the finger and lacks details on the 3D fingerprint ridge information. Since the ridge information in such images is essentially derived from the 2D images, it is adversely influenced from the variations in skin pigmentation, reflectance or its shape. Several studies have indicated that automatically locating the correspondence points in two fingerprint images, from different 3D views, is quite challenging and is another limiting factor for precise acquisition of 3D fingerprints.

2.2 Patterned Lighting

Another class of active imaging methods for live 3D fingerprint scans illuminate the fingers with *structured lighting* of known patterns, e.g. stripes, circles, etc., and measured deformation of the projected patterns to determine the 3D shape of presented finger. Such structured lighting based methods, unlike those in Fig. 2.2, acquire 3D fingerprint ridge details from their 3D geometry instead of their surface albedo. A faster and more versatile approach is to project many stripes, fringes or patterns once and simultaneously acquire multiple samples [9, 10]. Therefore, structured lighting patterns are often coded to automatically recover the correspondences between the projected pattern points and those observed from respective points in the acquired images. These correspondence points can be triangulated to recover 3D shape information.

A variety of structured lighting stripes or patterns have been investigated in the literature [9] and can be broadly categorized into one of the three categories, i.e. time-multiplexing (e.g. binary codes or phase shifting), spatial neighbourhood (e.g. De Bruijn sequences or M-arrays) and (c) colour coding (codification based on colour or grey levels). A comparative evaluation of representative techniques from each of three categories indicates [11] that the time-multiplexing techniques are easy to implement,

Fig. 2.2 Acquisition of 3D representation of fingerprint using multiple cameras

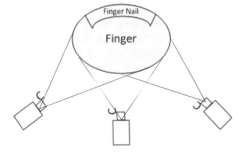

Fig. 2.3 Conceptual illustration of computing 3D fingerprint height using patterned lighting. The height of fingerprint z at a point (x, y) is proportional to the difference between measured phase ϕ_s and the reference phase ϕ_r, i.e. $z = s_{x, y} (\phi_s - \phi_r)$. The scaling factor $s_{x, y}$ is computed during the offline calibration

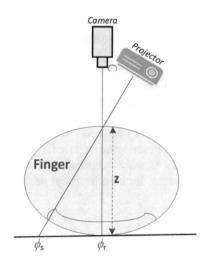

offers high spatial resolution and can provide more accurate depth details. Maximum resolution from such temporal coding techniques can be achieved by the combination of phase shifting and grey code but requires a large number of patterns or image frames that can pose problems for 3D fingerprint imaging due to involuntary motion of fingers. Therefore, 3D fingerprint imaging approaches implemented in [12, 13] uses phase measuring profilometry [14] (PMP) for encoding structural illumination to enable better precision with smaller number of image frames. These patterns are imaged from a calibrated camera which is synchronized with the projected patterns. The observed distortions in projected patterns from the camera images are computed by measuring their phase differences with reference as detailed in [12]. Figure 2.3 illustrates computation of 3D fingerprint height, at any specific point on 3D fingerprint surface, using calibrated phase difference measurements. The PMP approach projects sine-wave patterns whose phase is shifted several times. Traditional spatial phase unwrapping methods can accumulate errors from fingerprint ridge and valley discontinuities. Therefore, the implementation detailed in [13] computes the absolute phase pixel-by-pixel by incorporating optimum three-fringe number selection [15, 16] with a series of *colour* sinusoidal fringes.

Structured lighting based 3D fingerprint scanners have been commercially introduced [17] and offers one of the most promising solutions available today. Active 3D fingerprint imaging using coherent light source like laser or structured lighting can also enable liveness detection by analysing dynamic interference patterns or *biospeckle* as investigated in [18]. The projection and imaging of structured lighting patterns require time and, therefore, such methods can suffer from instability problems due to involuntary finger motions. In addition to the limitations with the reconstruction accuracy for the shape of high-frequency ridges, such methods generally require setup with higher cost and bulk, largely due to the requirement of a projector and high-speed camera.

2.3 Optical Coherence Tomography

Unlike popular methods of fingerprint imaging which are known to acquire finger skin surface ridge or dermal details, tomographic imaging of fingers can reveal inner skin layer between dermis and epidermis. Such 3D imaging of fingers can not only offer higher antispoofing capabilities [19] but can also enable recovery of damaged fingerprint ridges, due to scars or external factors, when dermis or surface-based 3D fingerprint acquisition is not viable. The OCT based 3D fingerprint imaging is essentially based on the principle of interefometery [20] where the light reflected from the finger skin, at various depth to recover subsurface 3D profile, is combined with the light reflected from a fixed mirror in an attached interferometer. Such an approach generally uses broadband near infrared illumination which can penetrate finger skin surface, or epidermis, to the subsurface layers. The interference patterns between light reflected by the fixed mirror and the finger surface layers are measured by a spectrometer. The depth information is retrieved by the analysis of spectral modulations in these patterns.

There have been many promising studies to acquire 3D fingerprints using OCT. Reference [21] reports recovery of small 4 mm × 4 mm × 2 mm sized 3D fingerprints. The work reported in the literature [22, 23] on the recovery of 3D fingerprints is quite preliminary as it acquires very small volume and under higher acquisition time which can pose challenges due to involuntary finger motion during the imaging. Large fingerprint imaging area is highly desirable as it can ensure adequate overlap between the 3D fingerprints acquired during the enrolment and subsequent verification stages. More recent efforts in [24] using full-field OCT enable the acquisition of 1.7 cm × 1.7 cm area but requires several seconds to build volumetric 3D fingerprint data from multitude of such images. In this context, the Fourier-domain OCT [25, 34], can acquire entire 3D fingerprint data in a single scan and offers much faster alternative but at a significantly higher cost. The cost of OCT-based sensors is prohibitively high, even full-field OCT sensor can currently cost over ten thousand US$, which limits its possible usage in popular practical fingerprint identification applications.

2.4 Ultrasound Imaging

Ultrasonic 3D imaging of fingerprints is based on *acoustic* time-of-flight measurements. The ultrasonic waves employed for such 3D fingerprint imaging can penetrate the epidermis layer of finger skin to enable acquisition of subsurface fingerprint details. Acoustic measurements from each point of the sensor/transducer are combined to reveal 3D fingerprint volume that can be horizontally sectioned to represent 2D fingerprint details at *different* subsurface levels.

The ultrasonic fingerprint sensors have been introduced for consumer applications more than a decade ago but were not popular, largely due to their relatively low-resolution images and bulk associated with the transducer. However, tremendous

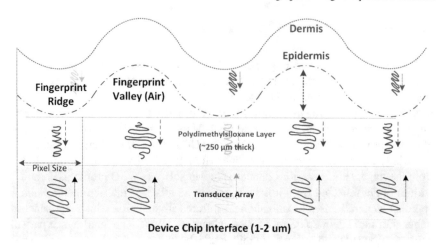

Fig. 2.4 Ultrasonic 3D fingerprint sensor interface using piezoelectric micromachined ultrasonic transducers

advances in microelectromechanical system technologies have addressed many of such constraints and enabled ultrasonic chips that are capable of acquiring sub-surface 3D fingerprint details. Ultrasonic 3D fingerprint imaging using capacitive micromachined ultrasonic transducers (CMUTs) has shown to offer 254 dpi resolution in [26]. More recent efforts using piezoelectric micromachined ultrasonic transducers (PMUTs) array have shown [27, 28] the development of a compact 3D fingerprint sensor with 500 dpi resolution. The acoustic signal in such PMUT array generates strong echo from the fingerprint valleys due to relatively large differences in impedance between the air and the sensor surface (Fig. 2.4). However, this difference in impedance for the echo from the fingerprint ridges is quite small. The acoustic waves in such 3D fingerprint sensors [27] penetrate into finger surface and are partially reflected from the dermis layer. It can be observed from Fig. 2.4 that such 3D fingerprint sensors are *not* completely contactless but offer quite promising advancement towards the development of low-cost 3D fingerprint sensors. A key limitation in such on-a-chip 3D fingerprint sensor is related to its smaller size (4.6 mm × 3.2 mm) which is more suitable for smartphone applications.

2.5   Photometric Stereo

Photometric stereo based approaches acquire multiple 2D images of the 3D finger surface under different illuminations but from a fixed viewpoint or using a fixed camera. The illuminators are located in such a way that the camera does not receive direct light from illuminators surrounding the camera. These illuminations are calibrated, i.e. their 3D spatial locations and resulting unit normal directions are computed during

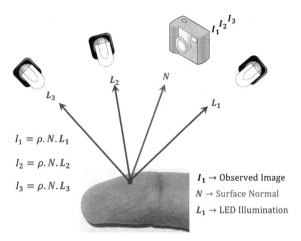

Fig. 2.5 Acquisition of 3D fingerprints using minimum/three 2D images under Lambertian surface assumption. A fixed camera acquires I_1, I_2, I_3 images, respectively, under L_1, L_2, L_3 illumination. The unknown surface normal N (n_x, n_y, n_z) at every pixel location is computed by solving three equations with three unknowns under the assumption of globally constant albedo/rho. The depth at every pixel points is computed from the integration of surface normals

offline steps. The changes in the observed grey-level intensity at every pixel locations in images (Fig. 2.5) depend on the 3D surface orientation and surface reflectance, in addition to the properties/locations of illumination. Therefore, by acquiring multiple images of 3D fingerprint, under each of the known or different illumination, a system of equations with relationships between unknown 3D surface orientation (along with surface reflectance) and known locations of fixed illuminations is generated. These equations are then solved to extract 3D surface orientation and surface reflectance information which is used to reconstruct 3D fingerprints surfaces and ridges. Recovery of 3D fingerprints using photometric stereo is sensitive to ambient illumination changes. Therefore, imaging setup must ensure that external illumination does not influence 3D fingerprint acquisition process. Acquisition of 3D fingerprints using photometric stereo can be achieved with a low-cost general purpose camera and fixed light-emitting diodes (LEDs). This approach is attractive due to its low cost and high precision that enables recovery of high-frequency ridge details. Therefore, acquisition of 3D fingerprints using photometric stereo is discussed in detail in the next chapter.

2.6 Other Methods

Low-cost, precise and fast acquisition of live 3D fingerprints is critical for the success of emerging 3D fingerprint identification technologies. Therefore, a range of other 3D imaging methods has been attempted in the literature to acquire live 3D fingerprint images. Shape from shading (SFS) is another attractive method, among shape from X techniques [29] that can be used to recover 3D fingerprint from a *single* 2D fingerprint image. This approach describes the 3D shape of a finger surface in terms of surface normal and recovers them from gradual intensity variations that are observed due to shading in a single 2D fingerprint image. Reference [30] details an improved SFS method for recovering 3D fingerprints by incorporating additional constraints on the brightness gradients and decomposing the acquired 2D fingerprint image into two frequency bands, which helps to minimize the errors in the recovery of smaller ridge details in 3D fingerprints. The SFS approach is highly sensitive to noise in the fingerprint image and its solution relies on surface continuity assumption, which can be very hard to meet due to nature of fingerprint ridges.

Another approach for the acquisition of 3D fingerprint detailed in [31] uses combination of fringe pattern projection and photometric stereo to recover 3D fingerprints. The pattern projection approach is used to reconstruct the structural part of the shading image while the texture or ridge patterns are recovered using cylindrical ridge model by considering them as semi-cylindrical structure. This investigation also uses a single fixed camera but the usage of a projector enhances the complexity of setup than those for the photometric stereo based 3D fingerprint acquisition. Time-of-flight (TOF) cameras [32] also offer another potential method of low-cost 3D fingerprint acquisition which is yet to be investigated in the literature. The TOF imaging approach employs a similar principle as for the 3D laser scanners. This approach computed the depth information from the phase shift between the (modulated) incident illumination and the reflected signal. The TOF imaging can acquire the depth information of the entire object simultaneously, unlike point-by-point scans for laser scanners, and can offer more attractive alternative to address involuntary finger motions during the 3D fingerprint image acquisition.

This chapter presented promises and limitations of live 3D fingerprint scan methods that have been introduced in the literature. Table 2.1 attempts to summarize emerging 3D fingerprint acquisition methods using the nature of imaging technique, acquisition mode, relative cost, expected reconstruction accuracy and also provides representative references. Acquisition of 3D fingerprints using laser-based range scanning [3] can address sensitivity of photometric method to the ambient illumination variations while acquiring absolute depth data. However, slow speed of such range scanning is the key limitations for its usage in 3D fingerprint acquisition. Photometric stereo and triangulation-based techniques are quite attractive for live 3D fingerprint imaging. The triangulation-based techniques can suffer from high-frequency noise but are known to offer robust reconstruction of low-frequency fingerprint surface shape. On the other hand, the photometric stereo based methods are known to perform very well in recovering high-frequency object details, such as the ridges structures required from 3D fingerprints. Therefore, next chapter includes a detailed explanation of this approach for the low-cost 3D fingerprint acquisition.

Table 2.1 Comparative summary of emerging 3D fingerprint image acquisition solutions

	Imaging principle	Source data	Acquisition mode	Relative cost[a]	Reconstruction accuracy[a]	Sample reference
Stereo camera	Triangulation	Range	Passive	Medium	Medium	[7]
Structured/patterned lighting	Triangulation	Range	Active	High	High	[12]
Photometric stereo	Shape from shading	Surface normal orientation	Active	Lowest	Very high	[33]
Optical coherence tomography	Interferometry	Backscattered light amplitude	Active	Very high	Very high	[21]
Ultrasonic imaging	Acoustic time of flight	Acoustic impedance	Active	Low	High-medium	[27]

[a]*Estimated*

References

1. Forensic Image Comparator3D (2018) http://www.sciencegl.com/fingerprint_3d/3D_AFIS.htm
2. Liu F, Zhang D (2013) Touchless multiview fingerprint acquisition and mosaicking. IEEE Trans Instrum Meas 62:2492–2502
3. Konica M, Non-contact 3D Digitizer Vivid 910/VI-910: instruction manual. Available online: http://www.konicaminolta.com/instruments/download/instructionmanual/3d/pdf/vivid-910vi-910instructioneng.pdf
4. The Hong Kong Polytechnic University Contact-free 3D/2D Hand Images Database (Ver. 1.0), http://www4.comp.polyu.edu.hk/~csajaykr/myhome/database_request/3dhand/Hand3D.htm
5. Hartley R, Zisserman A (2003) Multiple view geometry in computer vision, 2nd edn. Cambridge University Press, Cambridge, UK
6. Parizale G, Diaz-Santana E, Hauke R (2005) A multi-camera touchless device to acquire 3D rolled equivalent fingerprints. In: Advances in biometrics, LNCS 3832. Springer, Berlin, pp 244–250
7. Labati RD, Genovese A, Piuri V, Scotti F (2016) Towards unconstrained fingerprint recognition: a fully touchless 3D system based on two views on the move. IEEE Trans Syst Man Cybern 46:202–219
8. Fiumara G, Tabassi E et al (2018) Nail to nail fingerprint challenge, NIST IR 8210. https://doi.org/10.6028/NIST.IR.8210
9. Geng J (2011) Structured-light 3D surface imaging: a tutorial. Adv Opt Photonics 3:128–160. https://doi.org/10.1364/AOP.3.000128
10. Karpinsky N, Hoke M, Chen V, Zhang S (2013) High-resolution real-time 3D shape measurement on a portable device. In: Mechanical Engineering conference presentations, papers, and proceedings. Paper 67
11. Salvi J, Pages J, Battle J (2004) Pattern codification strategies in structured light systems. Pattern Recogn 37:827–849
12. Wang Y, Hassebrook LG, Lau DL (2010) Data acquisition and processing of 3-D fingerprints. In: IEEE transactions information forensics and security, Dec 2010, pp 750–760
13. Huang S, Zhang Z, Zhao Y, Dai J, Chen C, Xu Y, Zhang E, Xie L (2014) 3D fingerprint imaging system based on full-field fringe projection profilometry. Opt Lasers Eng 52:123–130
14. Yalla V, Hassebrook LG (2005) Very-high resolution 3D surface scanning using multi-frequency phase measuring profilometry. Proceedings of SPIE 5798, pp 44–53
15. Zhang ZH, Towers CE, Towers DP (2006) Time efficient colour fringe projection system for 3-D shape and colour using optimum 3-frequency interferometry. Opt Express 14:6444–6455
16. Towers CE, Towers DP, Jones JDC (2005) Absolute fringe order calculation using optimised multi-frequency selection in full-field porfilometry. Opt Lasers Eng 43:788–800
17. Flashscan3d, http://www.flashscan3d.com. Accessed May 2018
18. Chatterjee A, Bhatia V, Prakash S (2017) Anti-spoof touchless 3D fingerprint recognition system using single shot fringe projection and biospeckle analysis. Opt Lasers Eng 95:1–7
19. Cheng Y, Larin KV (2006) Artificial fingerprint recognition by using optical coherence tomography with autocorrelation analysis. Appl Opt 45:9238–9245
20. Gabai H, Shaked NT (2012) Dual-channel low-coherence interferometry and its application to quantitative phase imaging of fingerprints. Opt Express 20(24):26906–26912
21. Sousedik C, Breithaupt R, Busch C (2013) Volumetric fingerprint data analysis using optical coherence tomography. In: Proceedings of BIOSIG, Darmstadt, Sep 2013
22. Costa HSG, Bellon ORP, Silva L, Bowden AK (2016) Towards biometric identification using 3D epidermal and dermal fingerprints. In: Proceedings of ICIP, pp 3937–3941
23. Nehaus K, O'Gorman S, McNamara PM, Alexandrov S, Hogan J, Wilson C, Leahy MJ (2017) Simultaneous en-face imaging of multiple layers with multiple reference optical coherence tomography. J Biomed Opt 22
24. Anksorius E, Boccara AC (2017) Fast subsurface fingerprint imaging with full-field optical coherence tomography system equipped with a silicon camera. J Biomed Opt 22:096002

25. Wiser W, Bidermann BR, Klein T, Eigenwillig CM, Huber R (2010) Multi megahertz OCT: high quality 3D imaging at 20 million A-scans and 4.5 GVoxels per second. Opt Express 18:14685–14704
26. Lu Y, Tang H, Fung S, Wang Q, Tsai JM, Daneman M, Boser BE, Horsley DA (2015) Ultrasonic fingerprint sensor using a piezoelectric micromachined ultrasonic transducer array integrated with complementary metal oxide semiconductor electronics. Appl Phys Lett 106:263503
27. Tang H-Y, Lu Y, Assaderagh F, Daneman M, Jiang X, Lim M, Li X, Ng E, Singhal U, Tsai JM, Horsley DA, Boser BE (2016) 3D ultrasonic fingerprint sensor-on-a-chip. In: Proceedings of international solid state circuits conference ISSCC 2016, pp 202–204
28. Jiang X, Lu Y, Tang H-Y, Tsai JM, Ng EJ, Daneman MJ, Boser BE, Horsley DA (2017) Monolithic ultrasound fingerprint sensor. Microsyst Nanoeng 3:17059. https://doi.org/10.1038/micronano.2017.59
29. Buelthoff HH (1991) Shape-from-X: psychophysics and computation. In: Fibers' 91, SPIE, Boston, MA, pp 305–330
30. Balogiannis G, Yova D, Politopoulos K (2014) 3D Reconstruction of skin surface using an improved shape-from-shading technique. In: Proceedings of IFMBE, vol 41. Springer, pp 439–442. https://doi.org/10.1007/978-3-319-00846-2_109
31. Balogiannisa G, Yovaa D, Politopoulos K (2016) A novel non-contact single camera 3D fingerprint imaging system based on image decomposition and the cylindrical ridge model approximation. Int J Comput 20(1):174–198
32. Basler Time-of-Flight Camera, https://www.baslerweb.com/en/products/cameras/3d-cameras/time-of-flight-camera. May 2018
33. Kumar A, Kwong C (2013) Towards contactless, low-cost and accurate 3D fingerprint identification. In: Proceedings of CVPR, Portland, USA, June 2013, pp 3438–3443
34. Aum J, Kim J-H, Jeong J (2016) Live acquisition of internal fingerprint with automated detection of subsurface layers using OCT. IEEE Photonics Technol Lett 28:163–166
35. Liu F, Zhang D (2014) 3D fingerprint reconstruction system using feature correspondences and prior estimated finger model. Pattern Recogn 47:178–193

Chapter 3
Contactless and Live 3D Fingerprint Imaging

Online and precise acquisition of 3D finger images is one of the major technical challenges for the success of 3D fingerprint technologies. Photometric stereo based imaging offers low-cost approach to scan live 3D fingers and has also shown its effectiveness to reproduce high-frequency fingerprint ridge details. Therefore, detailed description of this approach is provided in this chapter.

3.1 Contactless 3D Finger Image Acquisition Using Photometric Stereo

Acquisition of 3D fingers using photometric stereo is based on the principle of appearance analysis of multiple contactless 2D finger images. These images are acquired under a calibrated setup using multiple (at least three) illumination sources fixed at different *non-coplanar* directions. These illumination sources are generally identical light-emitting diodes (LEDs) that emit light in the same spectrum. The reconstruction of 3D finger surface is based on the shape from shading technique. We assume that an orthographic camera, with large/infinite focal length, is used to acquire contactless 2D finger surface images. Given these 2D finger images $E(x, y)$, the shape from shading technique can be used to recover 3D finger surface $z(x, y)$ or simply the 3D fingerprint images. The surface reflectance R relates the observed 2D finger image pixel intensities in $E(x, y)$ to the surface orientations/gradients for a given source direction and the surface reflectance in the direction of imaging/camera as shown in the following:

$$E(x, y) = \rho I_0 R(p(x, y), q(x, y)) \tag{3.1}$$

where ρ is the albedo and I_0 is the incident radiance. The surface gradients $p(x, y)$ and $q(x, y)$ can be defined as follows:

© Springer Nature Switzerland AG 2018   29
A. Kumar, *Contactless 3D Fingerprint Identification*, Advances in Computer Vision
and Pattern Recognition, https://doi.org/10.1007/978-3-319-67681-4_3

$$p(x, y) = \partial f(x, y)/\partial x, \quad q(x, y) = (\partial f(x, y))/\partial y \tag{3.2}$$

Let us choose 3D coordinate system where the image plane coincides with x-y plane and the z-axis coincides with the viewing direction of fixed camera. The 3D finger surface can be reconstructed by recovering the surface height information $z = f(x, y)$. We approximate/consider finger surface as the Lambertian surface [1, 2] which is illuminated by multiple, say m, fixed light sources, i.e. $\mathbf{L} = [l^1, l^2, \dots l^m]^{\mathrm{T}}$. We also assume that light source and the camera are far away from the 3D finger surface to ensure consistency of illumination/viewing directions across the 3D finger surface. Each of these light sources (LED's) is fixed in an imaging device. The locations of each of these LED source direction $l = [l_x, l_y, l_z]^{\mathrm{T}}$, along with its radiance l, are known from the calibration stage. This calibration step is detailed in next section. Let $\mathbf{n} = [n_x, n_y, n_z]^{\mathrm{T}}$ be the *unknown* unit surface normal vectors at some point of interest on the 3D finger surface. The observed image pixel intensities $\mathbf{y}$, from the m 2D finger images, corresponding to the respective illumination sources can be written as follows:

$$\mathbf{y} = \mathbf{L}.\mathbf{x} \tag{3.3}$$

where $\mathbf{y} = [y_1, y_2, \dots y_m]^{\mathrm{T}}$ and $\mathbf{x} = \rho[n_x, n_y, n_z]^{\mathrm{T}}$. We assume that the light source directions are not coplanar so that the matrix $\mathbf{L}$ is non-singular. Equation (3.3) illustrates linear relationship between 3D finger surface, observed pixel intensities from 2D finger image and the unit surface normal vectors $\mathbf{x}$. The unknown vector $\mathbf{x}$ can be estimated from the standard [2] least squared error based solution technique using the following equation:

$$\mathbf{x} = \left(\mathbf{L}^T \mathbf{L}\right)^{-1} \mathbf{L}^T \mathbf{y} \equiv \rho \mathbf{n} \tag{3.4}$$

Since $\mathbf{n}$ is of unit length, the length of recovered vector $\mathbf{x}$ is absolute reflectance and represents the albedo ρ. The surface normal is represented by the direction of unit vector $\mathbf{n}$ obtained from (3.4). Therefore, the albedo map per pixel, along with its surface normal (n_x, n_y, n_z), for every 3D finger surface pixel can be recovered from (3.4) if the corresponding pixel can receive light from any of the LED source. The recovered surface normals are then integrated to recover the 3D fingerprint surface $z(x, y)$.

3.1.1 Imaging Setup and Calibration

Contactless 3D acquisition of 3D fingers using photometric stereo requires a fixed camera with multiple illuminators whose positions are *automatically* computed during the calibration process. Adequate illumination and non-coplanar positioning of illuminators are essential for the imaging process. Figure 3.1 illustrates the typical arrangement of illuminators (seven LEDs here) and fixed camera position in the cen-

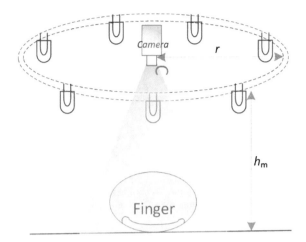

Fig. 3.1 Acquisition of 3D fingerprints using minimum/three 2D images under Lambertian surface assumption. A fixed camera acquires I_1, I_2, I_3

tre. The calibration process [3] can be fully automated and only the distance of LEDs from the imaging surface (h_m) is manually measured while they are fixed around a circle.

There are several approaches to automatically locate the positions of LEDs and a more simplified one is detailed in the following. We can use a small nail or pin to automatically identify the spatial location of different LEDs in the imaging setup. This nail is firstly placed somewhere in the middle of the *field of view* for the camera. We then illuminate each of the LED sources, seven in example case in Fig. 3.1, one by one and acquire one image for the correspondingly lit LED. These images will depict the expected shadow corresponding to the tail of nail and are shown in Fig. 3.2. Here, each of these images is of 768×1024 ($M \times N$) pixel size. The radius of circle for LEDs (r in Fig. 3.1) is 6.5 cm and the height (h_m) is 13.5 cm, while the pixel-to-centimetre ratio is 403/1. Let (x, y) represent the centre of circle where LEDs are symmetrically placed and (x', y') be the spatial position of the nail or pin placed on the imaging surface during the calibration.

The radius r and height h_m are firstly converted from centimetre scale to the pixel scale. Using each of the images in Fig. 3.2, we can measure positions of nail *tip* from the direction of accompanying shadow and these can be represented as (x_1, y_1), (x_2, y_2), ... (x_7, y_7). Next step is to transform the LED centre (x, y), pin centre (x', y') and each of the shadow/nail tip positions to the world coordinate system. Figure 3.3 details the conversion of image coordinates to the world coordinates during the calibration process.

It can be noted that the LED circle centre (x, y) and radius r form a circle while pin centre position (x', y') and shadow tip position (e.g. x_1, y_1) form a line. This can also be observed in Fig. 3.4 which shows respective line for one image in Fig. 3.2. Therefore, the position of LEDs is essentially the point of interaction with this *extended* line with the LED circle. In order to enhance the reliability in measurements, the nail is

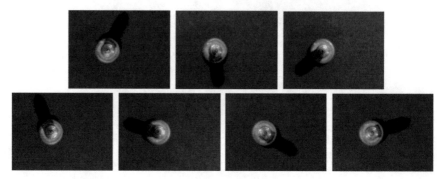

Fig. 3.2 Image samples acquired during the calibration process that represent the shadow from respective LED illumination in the imaging setup

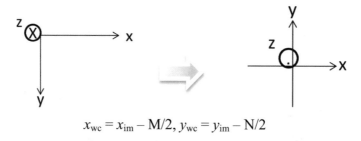

$$x_{\mathrm{wc}} = x_{\mathrm{im}} - \mathrm{M}/2, \ y_{\mathrm{wc}} = y_{\mathrm{im}} - \mathrm{N}/2$$

Fig. 3.3 Image coordinates $(x_{\mathrm{im}}, y_{\mathrm{im}})$ to world coordinate $(x_{\mathrm{wc}}, y_{\mathrm{wc}})$ conversion during calibration

placed at several different positions on the imaging surface and corresponding seven images (similar to as in Fig. 3.2) are acquired to compute the spatial position of each of the seven LEDs, in the same manner as discussed earlier. Figure 3.4a illustrates this interaction of extended line with LED circle while Fig. 3.4b illustrates the final locations of the intersection points that determine the location of LEDs. We can now use the height h_{m}, in pixel scale, to identify the spatial position of each of the seven LEDs in the *world coordinate* system.

Let us represent the position of such LEDs as posLx, posLy, posLz ($=h_{\mathrm{m}}$,) in world coordinate system. Next step is to transform and compute the pixel positions, corresponding to each pixel in the acquired image, into the image coordinate system. This transformation is achieved as follows:

$$\begin{aligned} \mathrm{pos}x &= x - (\mathrm{M}/2) - \mathrm{posL}x \\ \mathrm{pos}y &= y - (\mathrm{N}/2) - \mathrm{posL}y \\ \mathrm{pos}z &= \mathrm{posL}z \end{aligned} \tag{3.5}$$

where (M) and (N) represent the size of acquired image as discussed earlier. Each of the transformed pixel locations (pos$x$, pos$y$, pos$z$) are further normalized as follows:

$$\left[x', y', z'\right] = [x, y, z]/\sqrt{x^2 + y^2 + z^2} \tag{3.6}$$

(a)   **(b)**

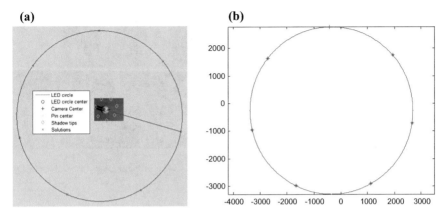

Fig. 3.4 **a** Localization of surrounded LEDs using the intersection of respective shadow line during the calibration and **b** Localized spatial positions of LEDs using average of multiple measurements

Fig. 3.5 Plain paper images acquired under different LEDs for the illumination normalization

The normalized pixel positions, for each of the LEDs, are stored and represent the key data from the calibration process. These positions represent L in Eq. (3.4) and are fixed for a given imaging setup. Therefore, $(L^T L)^{-1}$ can also be computed offline and retrieved/read during the online 3D finger imaging to compute the surface normal vectors.

The illumination received by different areas on the imaging surface, or from LEDs itself, can be different. Such uneven illumination from different LEDs can adversely influence the accuracy of surface normal estimation. Therefore, the illumination normalization is also performed during the calibration stage. Simplest approach for such illumination normalization is to acquire image from a plain white paper, under illumination from individual LEDs as shown in Fig. 3.5. The image normalization can be achieved as follows: $I_c(i,j) = I(i,j)/I_w(i,j)$, where I is the original image and

Fig. 3.6 Sequence of seven sample images acquired in quick succession under respective (LED)

I_c is the image after normalization, while I_w represents the corresponding image of the white paper. Such image illumination normalization step helps to ensure uniform illumination at all image points under any LED's illumination and is also part of offline calibration process.

3.1.2 Preprocessing Acquired 2D Images

The setup in Fig. 3.1 uses seven symmetrically distributed LED and a low-cost digital camera which can acquire 1280×1024 pixel images with 15 frames per second. Figure 3.6 shows seven such sample images acquired from a live 3D finger. Each of the live 3D fingers provides seven images and each under a different or unique LED illuminating the finger. Before locating the region of interest (ROI) for the reconstruction of 3D model, it is useful to combine all seven images and generate a *unified* or stacked greyscale image. The intensity of the resulting stacked image is normalized, i.e. $I_{all} = \left(\sum_{i=0}^{7} I_i\right)/\max\left(\sum_{i=0}^{7} I_i\right)$, where $\max(I_i)$ is the maximum intensity of the image. This stacking or combination of images with different LEDs is a linear operation, and the result is expected to be the same as a single image which is acquired under all (seven) simultaneously lit LEDs.

The stacked image is firstly subjected to edge detection operation, e.g. Canny, to localize the image boundaries. The resulting image is used to localize the boundaries of the finger. We scan the edge detected image to locate the first lines that overlap the edge lines from top and bottom in y-direction [4]. The average value of two lines plus and minus half of *predefined* ROI height will be the upper bound and lower bound of ROI. Simply tracing the upper bound and lower bound ROI lines from the right in x-direction, the position of the first edge pixel can be marked. The average of two positions in y-direction plus the offset serves as the right boundary of the ROI. The left boundary of the ROI is the right boundary minus the predefined ROI width.

Once the ROI is defined using the unified or the stacked image, it can be easily mapped to extract seven respective greyscale ROI images. Figure 3.7 illustrates such

Fig. 3.7 The ROI images automatically segmented from the images shown in Fig. 3.6. The size of these sample images is reduced from 2000×1400 to 500×350 pixels

automatically segmented ROI images corresponding to the acquired image samples shown in Fig. 3.6. The fingers presented for 3D imaging are frequently accompanied by sweat or oily/shiny contaminations which appear as specular reflections in acquired images. These specular reflections are considered as noise as they seriously degrade the accuracy of 3D reconstruction. Therefore, such pixels belonging to the specular reflections are identified and suppressed. Among many methods available to identify/suppress the specular pixels in the ROI images, more computationally efficient alternative is to reject those grey level, i.e. replace by zero, which are higher than a certain or predetermined threshold. The experiments using the setup or images in Fig. 3.6 also use such simplified but effective approach for suppressing the specular reflection. The grey-level pixel intensities from all seven images are firstly sorted or ranked in descending order. The top 0.228% pixel with high-intensity values in these seven images is ignored or considered as outliers belonging to specular reflections.

3.1.3 Surface Normal and Albedo

Multiple ROI images acquired from calibrated imaging setup are used to recover the surface normal vectors using least squared solution of (3.4), i.e. $\mathbf{x} = \left(\mathbf{L}^T \mathbf{L}\right)^{-1} \mathbf{L}^T \mathbf{y}$. In example case of images acquired using the setup shown in Fig. 3.1, $\mathbf{L} \in \mathbf{R}^{7 \times 3}$ is the pre-computed illumination direction matrix, $\mathbf{y} \in \mathbf{R}^{7 \times 1}$ represents the intensity or the vector representing grey levels observed at a given pixel position from each of the seven images and $\mathbf{x} \in \mathbf{R}^{3 \times 1}$ unknown surface normal vector at respective pixel positions. Surface reflectance or the unknown albedo for every 3D fingerprint pixel positions is recovered from the norm, i.e. $\rho = \text{norm}(\mathbf{x})$. The *unit* surface normal (n_x, n_y, n_z) for every pixel position is recovered from the norm, i.e. $\mathbf{n} = \mathbf{x}/\rho$. Figure 3.8 illustrates albedo and surface normal vectors recovered from a sample image. Recovered unit surface normal vectors are shown in red colour arrows on the albedo map.

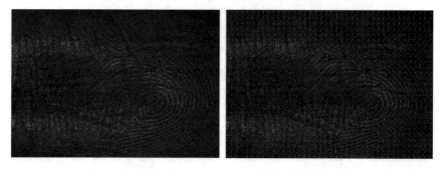

Fig. 3.8 Surface reflectance or albedo map (left) recovered from least squared solution for sample image and corresponding unit normal vectors (right) shown in red coloured arrows

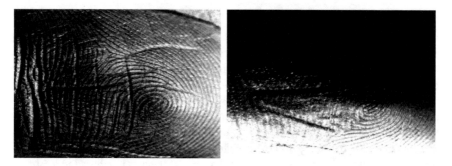

Fig. 3.9 Surface gradient $p(-n_x/n_z)$ on the left and $q(-n_y/n_z)$ for the sample in Fig. 3.8

At least three images are required to recover three unknown variables (n_x, n_y, n_z) in (3.4). Additional images, like seven images from the setup in Fig. 3.1, are however desirable to improve the robustness and enhance surface area coverage and accuracy of surface normal. The acquisition setup in Fig. 3.1 makes several standard assumptions for photometric stereo, i.e. the illumination source is at infinity and the camera is orthographic with its known parameters. The surface normal vectors extracted in Fig. 3.8 are employed next to compute surface gradients (p, q) along the x- and y-axes, respectively. Figure 3.9 illustrates such surface gradient map for the image sample shown in Fig. 3.8. It is quite known [5] that for a 3D surface represented by $z = f(x, y)$, the height z at a point (x, y) can be expressed in terms of the surface gradients, i.e. $p = z(x + 1, y) - z(x, y)$ and $q = z(x, y + 1) - z(x, y)$. Therefore, surface normal enables the recovery of 3D fingerprint depth map and is discussed in the next section.

3.1.4   Generating 3D Fingerprint Images

Surface gradients (p, q) and the reflectance (ρ) computed from the solution of (3.4) are used to recover cloud point data, which represents fingerprint height z in 3D space

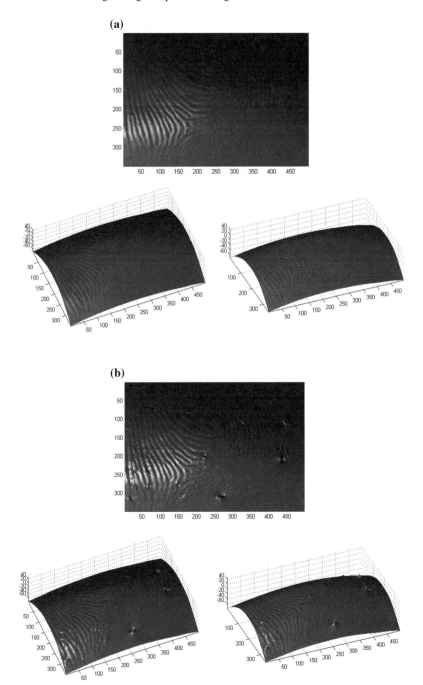

Fig. 3.10 Reconstructed sample 3D fingerprint surface using Frankot–Chellappa algorithm. Samples shown in **a** are without while those in **b** are with the usage normal amplification filter

and forms the raw 3D fingerprint data. There are several algorithms in the literature to reconstruct such 3D fingerprint cloud point data from the surface normals. This reconstruction operation is essentially the integration of gradient field recovered from the surface normal under the assumption that the recovered surface normals are continuous along the closed 3D surface. This assumption requires that the integral of gradient fields along any closed path should be equal to zero and such integration should be independent of the choice of integration path. The gradient fields recovered from real 3D finger surfaces are often non-integrable. This can be largely attributed to the noise in the estimation process, imaging errors or the assumptions made from the imaging setup in manipulation of surface normals.

Lack of consistency among the recovered surface normals is popularly known as the integrability problem [6] and several solutions are proposed in the literature. Most of these methods manipulate the recovered surface normals to achieve desired goal and the 3D image is recovered from the 2D integration of these manipulated surface normals. Such manipulation is achieved by incorporating *integrability constraints* to remove inherent ambiguities in the recovered surface normal or to regularize the desired solution. Online recovery of 3D fingerprint requires computationally simpler methods for the reconstruction, and therefore two methods, Frankot and Chellappa algorithm [7, 37] and Poisson Solver [6, 8], are briefly discussed for their usage. The implementations of these algorithms are simpler and also accessible [9, 10] in public domain.

Let the original 3D fingerprint surface required to be reconstructed, using the measured but non-integrable surface gradient field gradients (p, q) from (3.4), be represented by S_{3d}^p. Let us use (g_x^p, g_y^p) to represent the true gradient field of unknown 3D fingerprint surface S_{3d}^p. A common approach for the reconstruction of 3D surface is to minimize the least squared error $E(S_{3d}^p)$:

$$E\left(S_{3d}^p\right) = \iint \left(\left(g_x^p - p\right)^2 + \left(g_y^p - q\right)^2\right) dx dy \tag{3.7}$$

The Fourier transform of unknown 3D surface and Fourier transform of the gradients can be directly related using the Perseval's theorem [xx] as follows:

$$S_{3d}^p(u, v) = -j \frac{u G_x^p(u, v) + v G_y^p(u, v)}{u^2 + v^2} \tag{3.8}$$

where $G_x^p(u, v)$ and $G_y^p(u, v)$, respectively, represent the Fourier transform of gradient g_x^p and g_y^p. The Fourier transform of unknown 3D fingerprint surface S_{3d}^p is represented as $S_{3d}^p(u, v)$. It is straightforward now to recover the unknown (or integrable) 3D fingerprint surface gradients, or S_{3d}^p, by computing inverse Fourier transform of (8), i.e. $S_{3d}^p = F^{-1}\{S_{3d}^p(u, v)\}$. This approach is referred to as Frankot–Chellappa [7, 37] algorithm and has shown to be robust to noise [11] with several publicly accessible implementations [10]. Availability of Fast Fourier Transform implementations is also another reason for the popularity of this approach for online applications and is used here for the reconstruction of 3D fingerprints.

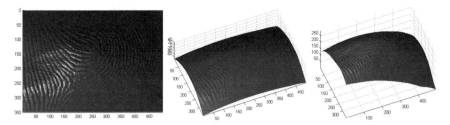

Fig. 3.11 Sample 3D fingerprint surface reconstructed using the Poisson solver

In order to ensure that surface gradients are integrable its curl along the closed path should be zero curl ($\nabla \times S_{3d}^p = 0$). Therefore, Euler–Lagrange relation, which gives the following Poisson equation offers another approach [8] to recover corrected or manipulated gradient fields.

$$\nabla^2 \left(g_x^p, g_y^p \right) = \mathrm{div}(p, q) \tag{3.9}$$

This approach is referred to as Poisson solver and detailed in [8] with public implementations in [10]. It offers another computationally simpler approach to reconstruct 3D fingerprint S_{3d}^p. Figure 3.10 shows sample 3D fingerprints reconstructed using the Frankot–Chellappa algorithm and the Poisson solver. These recovered surface normals can be further enhanced or amplified for their *visibility* as in [12]. Such amplification step firstly computes the average of surface normals over a local patch, say b. The difference between the actual surface normal and the computed average is amplified for the enhancement and this operation can be written as follows:

$$\tilde{\mathbf{n}} = \mathbf{n} + \alpha \left[\mathbf{n} - \frac{1}{bb} \sum_{i=1}^{b} \mathbf{n}_i \right] \tag{3.10}$$

The amplification step in (3.10) computes average of surface normals in a local $b \times b$ pixel region surrounding $\mathbf{n}$. The difference of this average from the actual surface normal is amplified by factor α and added to the surface normal $\mathbf{n}$ for the enhancement. It can be observed that the use of such amplification filter can help to increase the visibility of fingerprint ridges but it also introduces the odd points, i.e. random rapid changes in the height of reconstructed 3D fingerprints. It can also be observed from the results in Fig. 3.11 that the 3D fingerprints generated using the Poisson solver illustrate more natural shape.

3.1.4.1 Reconstruction Accuracy

Accurate acquisition of presented 3D fingers is critical for the success of contactless 3D fingerprint identification. However, lack of accurate 3D finger ground truth data

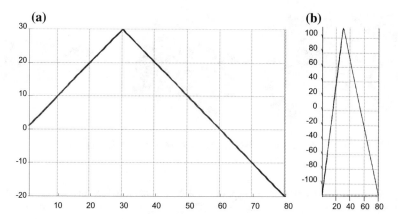

Fig. 3.12 Sample ground truth surface cross section **a** and the surface generated using Frankot–Chellappa algorithm **b**

makes it very difficult to evaluate the accuracy from a given imaging or reconstruction algorithm. Therefore, a known 3D surface model is generated as its cross section is shown in Fig. 3.12a. It can be observed from Fig. 3.12b that using Frankot–Chellappa algorithm, the slope of reconstructed surface has been changed to ensure continuity with the boundary points. The padding operation for surface normal vector field can reduce the distortion at boundary surface. However, it was observed that the reconstruction using Frankot–Chellappa algorithm is weak or susceptible to errors at discontinuous points, i.e. near the edges of 3D fingerprint or ROI. The ground truth model for the evaluation of 3D reconstruction is currently obtained from a laser scanner [13]. A 3DMD scanner [14], which is popular for medical imaging applications, with 0.5 mm RMS was employed. There are ranges of industrial laser scanners which can offer much higher accuracy (up to 0.001 inch or 0.00254 cm). These can be employed to generate ground truth 3D fingerprint data and used to ascertain accuracy of reconstructed 3D fingerprints.

3.1.4.2 Imaging Resolution

In order to measure the resolution of the reconstructed 3D fingerprints using the setup in Fig. 3.1, we place a ruler at 1.2 cm above the imaging surface/ground and this distance can represent the average height of the 3D finger surface. We choose two points in the acquired image of this ruler, which are 2137 pixels apart. The real-world or actual distance between these points is 2 cm, and therefore the spatial resolution is 9.359 μm (2 cm/2137 pixel). The minimum resolvable separation or MRS $\approx \frac{1.22\lambda d_0}{D}$, where d_0 is the average distance between the 3D finger surface and the lens, D is the diameter of the lens and λ is the wavelength for the peak of LED illumination. For the imaging system in Fig. 3.1, $d_0 = 9.8$ cm and $D = 2.5$ cm, and can be approximated for the blue light ($\lambda = 550$ nm) used during the imaging which also has shorter wavelength. The approximated MRS is therefore 2.630 μm. It should be noted that for a fixed camera-based imaging system, the image resolution is not constant over the imaged area. The PIV standards [15] introduced from FBI are widely used to evaluate the capabilities of commercial 2D fingerprint sensors. These standards mandate the use of calibration targets, sine gratings on a flat surface, to facilitate the measurements of quantitative imaging thresholds from fingerprint sensors. In medical imaging applications, like for computerized tomography or radiography, the 3D targets or phantoms that represents the characteristics of expected imaging targets during the deployments are widely employed. Similar method has been detailed in reference [16] to evaluate contactless 2D fingerprint sensors using 3D cylindrical targets that are mapped with 2D fingerprint ridge—valley patterns. Fabrication of such 3D fingerprint targets [17] using the materials that closely resemble the human fingers, i.e. ridge–valley characteristics and surface reflectance, can be used to evaluate the fidelity of 3D fingerprint images. Availability of commercial 3D fingerprint sensors in coming years is expected to initiate the standardization and development of 3D fingerprint targets in coming years.

3.1.5 Removing Specular Reflection

Finger skin surface, like many other non-lambertian surfaces, is often accompanied by sweat, oily or greasy contaminations which generate specular reflections in the contactless 2D images acquired for the 3D reconstruction. Such specular reflection from unwanted contaminations generates noise and degrades the accuracy of reconstructed 3D fingerprints. Therefore, it is desirable to suppress or eliminate such specular pixels in the 2D images before the reconstruction. Hierarchical selection of Lambertian reflectance [18] is one of the more effective methods to automatically identify these specular pixels and is briefly discussed. Let us assume that the vector $Is = \{I_1, I_2...I_m\}$, with $m \geq 3$, represents 2D fingerprint ROI pixels. In 3D spaces, any four (illumination) vectors are linearly dependent and therefore we can write

$$a_1 I_1 + a_2 I_2 + a_3 I_3 + a_4 I_4 = 0 \tag{3.11}$$

for some real coefficients a_k with $k = 1,...,4$. Both sides of the above equation can be multiplied by surface normal vector and albedo to generalize the Lambertian error [19]. In summary, the key idea is that the four vectors or pixels are linearly dependent or related such that the coefficients in above equation can be computed from the zero-sum relationship. For specific example of imaging setup in Fig. 3.1 with seven illumination sources ($m = 7$), this error can be written as follows:

$$E_L = \sum_{i=0}^{\frac{7!}{4!3!}} \sum_{k=1}^{7} a_{ik} I_{S_k} \tag{3.12}$$

(a) **(b)**

(c)

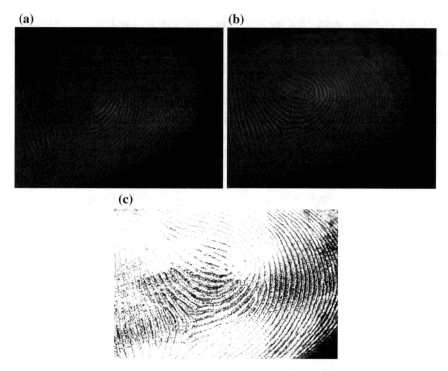

Fig. 3.13 Sample images **a–b** with specular pixels acquired from setup in Fig. 3.1 and the image **c** representing the mapping of specular light affected pixels

which is the sum of error for every combination of selecting four vectors in I_{Sk} from the seven vectors. When E_L is larger than a threshold, the highest value in I_{Sk} can be considered as the specular light/pixel and is ignored, i.e. we use the rest six darker pixel values in the computations. Figure 3.13 shows sample contactless fingerprint images with specular reflections and sample results generated from this approach for identifying specular pixels.

This approach [18] is computationally complex as it requires us to compute solutions of (3.11) for each of the presented 3D fingerprints. Our observations have indicated that this approach generates slightly better results for the reconstructed 3D fingerprints. However, such marginal improvement is not sufficient to justify computational requirements for online system, and therefore we preferred a simpler approach to identify/suppress specular reflection. We top K percentage of high-intensity pixels from all the 2D fingerprint ROI images, with different illumination profiles which are acquired for a 3D fingerprint reconstruction, as the specular reflections. The magnitude of K is empirically determined and was fixed as 0.228 during all our experiments (Figs. 3.10 and 3.11).

3.1.6 Addressing Non-Lambertian Influences During 3D Fingerprint Imaging

The real-world 3D fingerprint source image data is not expected to comply with the Lambertian surface reflectance requirements in (3.1) the formation of images. Such non-Lambertian effects can be attributed to the specular reflection, shadows, sensor/source noise or importantly due to the nature of finger skin which can be largely considered as a translucent surface. In order to address such non-Lambertian influences on classical photometric stereo method, several approaches have been introduced in the literature. These approaches can be broadly categorized into two classes and are briefly introduced in the following two sections.

3.1.6.1 Statistical Methods

The first category of approaches is those, which use statistical methods that attempt to eliminate the adverse influence from the non-Lambertian effects. Given the illumination matrix $\mathbf{L}$ in (3.3), acquired during the calibration of setup in Fig. 3.1, the photometric stereo (PS) problem can be formulated as the solution to following optimization problem:

$$\underbrace{\min}_{b_j} \left\| \mathbf{y}_j - \mathbf{L}^T \mathbf{b}_j \right\|^2 \tag{3.13}$$

where $\mathbf{b}_j = \rho\mathbf{n}$ and $j = 1, 2, \dots m$. In previous section, we employed simplest solution for the above equation using least square (LS) error approximation solution [20]. There are several other promising methods proposed in the literature for the optimization problem in (3.11). Wu et al. [21] explicitly add sparse corruptions to the optimization. They formulate the optimization problem as the problem for recovering a low-rank matrix. Ikehata et al. [22] extended this approach using a rank-3 Lambertian constraint. Figure 3.14 provides comparative results for recovered surface normal vectors, using the images acquired from the setup in Fig. 3.1, between the SBL method detailed in [5] and the classical LS method. Slight improvement in the accuracy of recovered surface normal is at the expense of enhanced computational requirements which may not be attractive for online 3D fingerprint reconstruction.

The statistical methods for optimization problem in (3.13) generally offer superior performance when large a number of input images are employed. However, when the number of such images is smaller, say smaller than five, the achievable performance is similar or worse than that can be achieved using classical Lambertian method [2, 20]. In addition, when the number of input images is bare minimum or three, most of these methods [5, 21, 22] perform poorly. With the availability of large number of input images, the optimization problem can be expected to be solved more accurately. Acquiring smaller number of images, to reconstruct 3D fingerprints, is Fig. 3.1 highly desirable to prevent adverse impact from involuntary finger motion and to reduce the

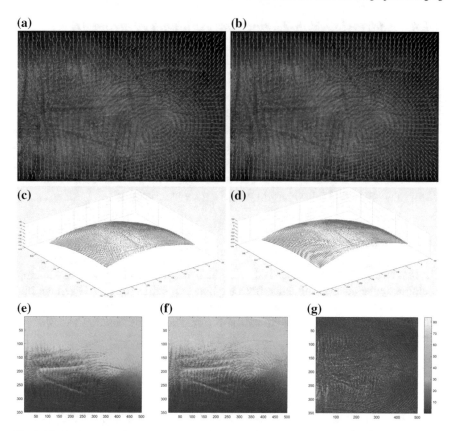

Fig. 3.14 Surface normal orientations (blue arrows) recovered from a sample 3D fingerprint using **a** classical PS method, and **b** SBL method. Reconstructed 3D fingerprint from different views using **c** classical PS method and **d** SBL method. Heat map of recovered surface normal from classical PS method in **e**, SBL method in **f** and their difference in **g**. The mean angular difference between the surface normal orientations in ∈ and **f** is 9.6°

complexity. Therefore, reference [23] revisits the photometric stereo problem under non-Lambertian reflectance assumption and derives a new optimization formulation for real objects. This approach attempts to minimize the error between the pixel intensity from the real-world illumination reflectance and the corresponding Lambertian reflectance. Figure 3.15 illustrates comparative experimental results from a 3D fingerprint sample acquired using setup shown in Fig. 3.1.

The images in Figs. 3.15, 3.16a, b and c, respectively, illustrate reconstructed results from the two different samples using the imaging setup in Fig. 3.1. In each of these two figures, the first column of images represents the results from LS method using all seven images, the second column of images represents the results from the method in [23] using *three* images and the third row of images represents the results from LS method using *three* images. Two different rows in each of these figures provide two different views of the reconstructed result. It can also be ascertained

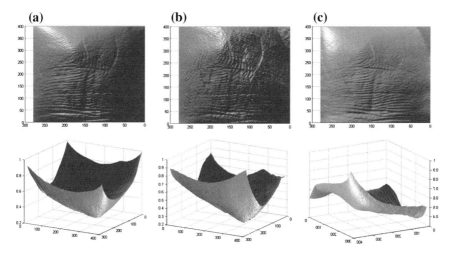

Fig. 3.15 Sample results from fingerprint reconstruction using LS method: **a** all seven images, **b** method in [23] using only three images and **c** LS method using only three images

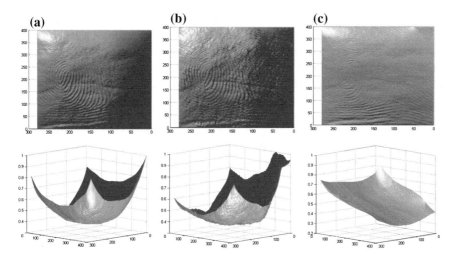

Fig. 3.16 Sample results from fingerprint reconstruction using LS method: **a** all seven images, **b** method in [23] using only three images and **c** LS method using only three images

from the image samples in Figs. 3.15 and 3.16 that the results using the method in [23] offer better details in recovering fingerprint ridges than those from using the LS method.

3.1.6.2   Non-Lambertian Surface Reflectance Based Methods

The second category of methods uses more sophisticated or realistic surface *reflectance* model, to account for the non-Lambertian characteristics observed in real-world object materials. Georghiades [24] adopted the Torrance–Sparrow model [25] to address this problem. Reference [26, 27] assumes specular-spike reflectance and provided solutions by detecting specular spots in the acquired images. Reference [28] introduces theory of photometric surface reconstruction from image derivatives and recover surface information from lower order derivatives. Among several other methods to reconstruct non-Lambertian surfaces using photometric stereo, the Hanrahan–Krueger (HK) model [29] attempts to account for the influences from the multiple scattering under multilayered structure of skin. Finger skin surface exhibits such translucent and non-Lambertian surface characteristics. Therefore, this model is quite attractive to recover 3D fingerprint surface and merits discussion.

Recovering 3D Fingerprints Using HK Model

The HK model assumes that the light reflected by finger skin, or any material that exhibits subsurface scattering, can be computed from the arithmetic combination of incident light reflected by each layer multiplied by the percentage of incident light that actually reaches the respective layer. The finger skin is modelled as a two-layer surface, consisting of the epidermis and the dermis, each with different reflectance properties. Figure 3.17 illustrates how the incident light L_i is reflected and the components of reflected light L_r from different subsurface layers. In the original HK model [29], each layer is parameterized by the absorption cross section σ_a, the scattering cross section σ_s, the thickness of the layer τ_d, g is the mean cosine value of the phase function and φ represents the angle between the incident light direction and the view direction. The intensity of scattered light from a subsurface layer (Fig. 3.17), using this model, can be defined [29] as follows:

$$L_{r,v} = \frac{\sigma_q}{\sigma_t} \frac{\left(1 - g^2\right)}{\left(1 + g^2 - 2g\ \cos\varphi\right)^{\frac{3}{2}}} \frac{\cos\theta_i}{(\cos\theta_i + \cos\theta_r)} \left(1 - e^{-\sigma_t d\left(\frac{\cos\theta_i + \cos\theta_r}{\cos\theta_i \cos\theta_r}\right)}\right) + \rho \cos\theta_i$$

$$= L_{r,v}^1 + \rho \cos\theta_i \tag{3.14}$$

where $L_{r,v}^1$ represents the first-order scattering term and $\rho \cos\theta_i$ is the approximation for higher order terms ($L_{r,h}$). Above equation uses simplified HK model where only the response from epidermis layer is accounted while the scattering in dermis layer is ignored. This approximation reduces the contributions in the final image but further attempts to compute too many unknown parameters can degrade the estimation accuracy of main parameters.

$$\sigma_q = \frac{L_i \sigma_s T^{21} T^{12}}{4\pi}, \quad \sigma_t = \sigma_s + \sigma_a, \zeta = \frac{\sigma_s}{\sigma_t} \tag{3.15}$$

$$L_r = L_{r,s} + L_{r,v} \tag{3.16}$$

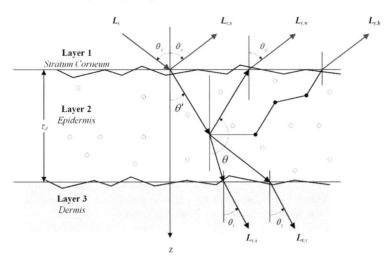

Fig. 3.17 Modelling scattering, reflection and refraction from finger skin under non-Lambertian assumption using HK model

Assuming $L_{r,s}$ is eliminated by rejecting the specular pixel intensity, the observed intensity L_r, is largely depended on $L_{r,v}$. The T^{12} and T^{21} are the Fresnel transmittance coefficients for the forward and backward directions, respectively, and can be assumed to be constant over the whole surface. We can formulate the following non-linear optimization problem to recover unknown parameters for the simplified HK model.

$$\text{Optimization problem}: \arg \min E(x_j), \text{ where } E(x_j) = \sum_{k=1}^{M} \left(L_r^k - I^k\right)^2 \quad (3.17)$$

where M is the number of valid input pixel intensity and I represents grey level of pixels from each of the observed image. Using [29], we can choose $\{\tau_d, \sigma_s, \sigma_a, g\}$ as $\{0.12 \text{ mm}, 30 \text{ mm}^{-1}, 4.5 \text{ mm}^{-1}, 0.81\}$, respectively. Similarly, from details in [12, 30], σ_s and σ_a can be fixed as 6 and 8.5 mm^{-1} for the blue colour light which peaks at 470 nm. The depth of epidermis can be assumed to be 0.325 mm as in [31]. Therefore, $\{\tau_d, \sigma_s, \sigma_a, g\}$ can, respectively, $\{0.325 \text{ mm}, 6 \text{ mm}^{-1}, 8.5 \text{ mm}^{-1}, 0.81\}$ be another set of fixed parameters for our investigation. Assuming T^{12} and T^{21} as fixed parameters, σ_q is then proportional to T^{12} and T^{21}. The set of *unknown* parameter vector in (3.17) can be consolidated in a vector $x_j = \left(n_x^j, n_y^j, n_z^j, \rho^j, \sigma_q^j\right)$ and its transformation to the spherical coordinate system can be represented as $x_j = \left(\theta^j, \phi^j, \rho^j, \sigma_q^j\right)$, where $\theta = \arcsin(n_z)$ and $\phi = \arctan(n_y/n_x)$. In order to solve optimization problem in (3.17), it is desirable to select a robust optimization method, which can converge efficiently. The method employed in our experimentation was based on the unconstrained Newton method, with additional functions/constraints to ensure that ρ and σ_q remain positive at the end of every iteration.

- $$\rho > 0, \sigma_q > 0 \qquad (\sigma_s, T^{12}, T^{21}, L_i \text{ are positive})$$

- $$0 < \theta \le 90 \qquad \text{(all surface pixels are visible to camera)}$$

- If $\phi \ge 360$ then $\phi = \phi - 360$
- If $\phi < 0$ then $\phi = \phi + 360$

It should be noted that $\cos\theta_i = \vec{l}\cdot\vec{n}$, $\cos\theta_r = \vec{r}\cdot\vec{n}$ and $\cos\varphi = -\vec{l}\cdot\vec{r}$ for the backward scattering. Let n_1 and n_2 be the refractive indices of medium 1 and medium 2. When the LED light enters perpendicularly on the surface of the medium 2 from the medium 1, the specular reflectance R_0 can be computed as follows:

$$R_0 = \frac{(n_1 - n_2)^2}{(n_1 + n_2)^2}, \tag{3.18}$$

where the refractive index n is 1.0 for the air and 1.38 for the epidermis [32]. Using Schlick's approximation in [33, 34], the Fresnel reflection factor R can be computed as follows:

$$R = R_0 + (1 - R_0)(1 - \cos\theta_i)^5 \quad \text{for } n_1 < n_2 \tag{3.19}$$
$$R = R_0 + (1 - R_0)(1 - \cos\theta_r)^5 \quad \text{for } n_1 > n_2 \tag{3.20}$$

The reflected energy coefficient R_p is defined as

$$R_p = |R|^2 \tag{3.21}$$

It is known from the total energy conservation principle that $R_p + T_p = 1$. The product of T^{12} and T^{21} can be computed as in the following:

$$T^{12} = \frac{n_2^2}{n_1^2}(1 - R_{p12}) \tag{3.22}$$

$$T^{21} = \frac{n_1^2}{n_2^2}(1 - R_{p21}) \tag{3.23}$$

$$\begin{aligned} T^{12}T^{21} &= (1 - R_{p12})(1 - R_{p21}) \\ &= \left(1 - (R_0 + (1 - R_0)(1 - \cos\theta_i)^5)^2\right)\left(1 - (R_0 + (1 - R_0)(1 - \cos\theta_r)^5)^2\right) \end{aligned} \tag{3.24}$$

The use of HK model to recover 3D fingerprint surface is detailed in [12], also employed in [36], with promising results. Therefore, such attempts were investigated and sample comparative results are reproduced in Fig. 3.18. Our overall attempts from multiple 3D fingerprint samples indicated that the results from Lambertian model and the HK model are very similar.

One possible indicator of comparative performance is the energy, as also employed in (3.17), and comparative numbers for two sample images are shown in Table 3.1. The second sample results in this table are from the images shown in Fig. 3.18. The results from HK model illustrate smaller energy but smaller energy is not an indicator for better 3D reconstruction results. For the spatial points with poor conditions, e.g. specular points and shadowed points, the Lambertian model generally illustrated higher energy than the HK model but it is closer to the real value (Fig. 3.19). Despite lack of comparatively superior results using HK model, such approach offers lot of promises and underlines the need for further work. The key shortcoming in the use of HK model in our work is plausibly in the consideration of 3D geometry of fingerprint surface. Further work is required to account for the ridge–valley 3D geometry of finger skin and higher order subsurface scattering terms in (3.14) which were ignored in the experiments.

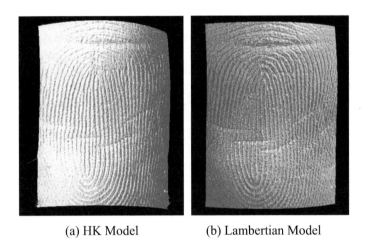

(a) HK Model   (b) Lambertian Model

Fig. 3.18 Comparative results from same finger using HK model and Lambertian model

Table 3.1 Total energy from the two methods for two sample fingerprint images

	Fingerprint sample 1	Fingerprint sample 2
Lambertian model	8.3246×10^9	2.7087×10^7
HK model	8.3237×10^9	2.7068×10^7

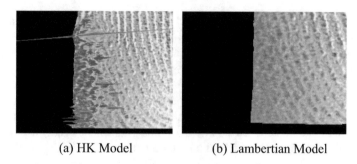

(a) HK Model   (b) Lambertian Model

Fig. 3.19 Reconstruction results from shadowed region: **a** results using HK model and **b** results using Lambertian model

3.2 Complexity for Online 3D Fingerprint Acquisition

Computational complexity of contactless 3D fingerprint image acquisition largely depends on the size of images. Table 3.2 details average computational time for the recovering surface normal and albedo (using least squared solution) for images of various sizes expressed in pixels. Similarly, Table 3.3 details the average computational time for recovering depth details of 3D fingerprints using Frankot–Chellappa algorithm. Total computational time for 3D reconstruction is the sum of respective time in Tables 3.2 and 3.3. These computational times are computed for a system implemented in Windows 10 Pro operating system and run on a computer with i5-7200 CPU@2.50 GHZ, 2 cores, 4 logical processors. It is worth mentioning that the computational time for large image size (2000 × 1400) in these tables is less reliable as the paging and memory of the system was only 4 GB size, and therefore system with larger memory should be used to ensure that swapping of memory pages is completely avoided. The implementation speed detailed in Tables 3.2 and 3.3 have been improved with the usage of faster library FFTW [35] and pre-computing the fixed term $\left(\mathbf{L}^T\mathbf{L}\right)^{-1}\mathbf{L}^T$ for least squared solutions.

Table 3.2 Complexity for recovering 3D fingerprint surface normal and albedo

Image size	Execution time (s)
2000 × 1400	3.45
675 × 519	0.43
225 × 169	0.047

Table 3.3 Complexity for recovering 3D fingerprint depth using Frankot–Chellappa algorithm

Image size	Execution time (s)
2000 × 1400	4.11
675 × 519	0.10
225 × 169	0.03

References

1. Horn B (1990) Height and gradient from shading. Int J Comput Vision 5:37–75
2. Kumar A, Kwong C (2015) Towards contactless, low-cost and accurate 3D fingerprint identification. IEEE Trans Patt Analy Mach Intell 37:681–696
3. Zhang Z (2000) A flexible new technique for camera calibration. IEEE Trans Pattern Anal Mach Intell 22:1330–1334
4. Kumar A, Kwong C (2012) Contactless 3D biometric feature identification system and method thereof. US Provisional application No. 61/680,716
5. Ikehata S, Wipf D, Matsushita Y, Aizawa K (2012) "Robust photometric stereo using sparse regression. In: Proceedings of CVPR 2012, Providence, USA, pp 318–325
6. Agrawal A, Raskar R Chellappa R (2006) What is the range of surface reconstructions from a gradient field? In: Proceedings of ECCV 2006, Graz, Austria
7. Frankot RT, Chellappa R (1987) A method for enforcing integrability in shape from. Proc Int. Conf., Computer Vision, ICCV
8. Simchony T, Chellappa R, Shao M (1990) Direct analytical methods for solving poisson equations in computer vision problems. IEEE Trans Pattern Anal Machine Intell 435–446
9. http://www.peterkovesi.com/matlabfns/Shapelet/frankotchellappa.m. May 2018
10. http://www.amitkagrawal.com/eccv06/RangeofSurfaceReconstructions.html
11. Schlüns K, Klette R (1997) Local and global integration of discrete vector fields. In: Solina F, Kropatsch WG, Klette R, Bajcsy R (eds) Advances in computer vision. Advances in computing science. Springer, Vienna
12. Xie W, Song Z, Zhang X (2010) A novel photometric method for real-time 3D reconstruction of fingerprint. In: Proceedings international symposium on visual computing LNCS 6454, Springer, pp 31–40
13. Seitz S, Curless B, Diebel J, Scharstein D, Szeliski R (2006) A comparison and evaluation of multi-view stereo reconstruction algorithms. Proc CVPR 2006:519–526
14. http://www.dirdim.com/prod_laserscanners.htm. Accessed May 2018
15. Personal Identity Verification (PIV) of Federal Employees and Contractors, US Department of Commerce, FIPS PUB 201–2, Aug. 2013, https://nvlpubs.nist.gov/nistpubs/FIPS/NIST.FIPS.201-2.pdf
16. Orandi S, Byers F, Harvey S, Garris M, Wood S, Libert JM, Wu JC (2016) Standard calibration target for contactless fingerprint scanners, US Patent No. 9,349,033 B2
17. Engelsma J, Arora SS, Jain AK, Paulter NG (2018) Universal 3D wearable fingerprint targets: advancing fingerprint reader evaluations. IEEE Trans Info Forensics Security 13(6):1564–1578
18. Bringer B, Bony A, Khoudeir M (2012) Specularity and shadow detection for the multisource photometric reconstruction of a texture surface. J Opt Soc Am A 29:11–21
19. Barsky S, Petrou M (2003) The 4-source photometric stereo technique for three-dimensional surfaces in the presence of highlights and shadows. IEEE Trans Patt Analy Mach Intell 25:1230–1252
20. Woodham RJ (1994) Gradient and curvature from photometric stereo including local confidence estimation. J Opt Soc America 11:3050–3068
21. Wu L, Ganesh A, Shi B, Matsushita Y, Wang Y, Ma Y (2011) Robust photometric stereo via low-rank matrix completion and recovery. In: Proceedings of ACCV 2010, Springer, pp 703–717
22. Ikehata S, Wipf D, Matsushita Y, Aizawa K (2012) Robust photometric stereo using sparse regression. Proc CVPR 2012:318–325
23. Zheng Q, Kumar A, Pan G (2015) On accurate recovery of 3D surface normal using minimum 2D images. Technical Report No. COMP-K-20. http://www.comp.polyu.edu.hk/~csajaykr/COMP-K-20.pdf
24. Georghiades AS (2003) Incorporating the torrance and sparrow model of reflectance in uncalibrated photometric stereo. Proc ICCV 2003:816–823
25. Torrance KE, Sparrow EM (1967) Theory for off-specular reflection from roughened surfaces. J Opt Soc Am 57(9):1105–1112

26. Drbohlav O, Chaniler M (2005) Can two specular pixels calibrate photometric stereo? Proc ICCV 2005:1850–1857
27. Drbohlav OR, Šára R (2002) Specularities reduce ambiguity of uncalibrated photometric stereo. In: Proceedings of ECCV 2002, Springer, pp 46–60
28. Chandraker M, Bai J, Ramamoorthi R (2013) On differential photometric reconstruction for unknown, isotropic brdfs. IEEE Trans Patt Anal Mach, Intell 35:2941–2955
29. Hanrahan P, Krueger W (1993) Reflection from layered surfaces due to subsurface scattering. In: Proceedings of SIGGARPH, pp 165–174
30. Li L, Ng CS-L (2009) Rendering human skin using a multi-layer reflection model. Int J Mathematics Comput Simul 3:44–53
31. Jacques SL (1989) Skin optics. Oregon Medical Laser Center News, https://omlc.org/news/jan98/skinoptics.html. Accessed July 2018
32. Weyrich T, Matusik W, Pfister H, Bickel B, Donner C, Tu C, McAndless J, Lee J, Ngan A, Jensen HW Gross M (2006) Analysis of human faces using a measurement-based skin reflectance model. In: Proceedings of SIGGRAPH 2006, Boston, pp 1013–1024
33. Schlick C (1994) An inexpensive BRDF model for physically-based rendering. Comput Graphics Forum 13:233–246
34. Greve BD (2006) Reflections and refractions in ray tracing, https://graphics.stanford.edu/courses/cs148-10-summer/docs/2006–degreve–reflection_refraction.pdf
35. FFTW (2018) C subroutine library for discrete Fourier transform, http://www.fftw.org
36. Pang X, Song Z, Xie W (2013) Extracting valley-ridge lines from point-cloud-based 3D fingerprint models. IEEE Comput Graphics Appl 73–81
37. Frankot RT, Chellappa R (1988) A method for enforcing integrability in shape from shading Algorithms. IEEE Trans Pattern Anal Machine Intell 439–451

Chapter 4
3D Fingerprint Acquisition Using Coloured Photometric Stereo

Acquisition of contactless 3D fingerprint images has been discussed in the previous chapter and offers low-cost alternative to several other methods summarized in Chap. 2. As also underlined in [1], the strength of photometric stereo based 3D surface reconstruction lies in accurately reproducing high-frequency information. Therefore, this approach is quite attractive in reproducing 3D ridge–valley structure in fingerprints and such accuracy is critical for the success of contactless 3D fingerprint identification. However, involuntary finger motion during the acquisition of multiple contactless 2D finger images, required to reconstruct 3D fingerprints using photometric stereo, can pose serious challenges as fingers are expected to be standstill during such multiple imaging shots. Therefore, this chapter details the development of more effective 3D fingerprint acquisition using coloured (Red, Green and Blue) LEDs and a fixed camera. Unlike the method in the previous chapter, which incorporated seven images, only two image shots or at least one shot is employed to reconstruct 3D fingerprint from the presented fingers. This approach [2] also incorporates automated detection of involuntary finger motion and uses such detection to improve recovery of 3D fingerprints. Such an approach results in significantly faster acquisition of 3D fingerprint images and is discussed in the following.

4.1 Image Acquisition Setup for Coloured 3D Photometric Stereo

This image acquisition setup uses a single fixed 2D camera and six coloured LEDs, i.e., two blue, two green and two red LEDs, respectively. These LEDs are *symmetrically* distributed, around the camera lens as shown in Fig. 4.1. It is ensured that the camera lens does not receive direct illumination from any of the LEDs. The average distance between the camera and finger in this setup was about 75 mm. The acquisition of fingerprint image is synchronized with the switching of LED illumination.

© Springer Nature Switzerland AG 2018

A. Kumar, *Contactless 3D Fingerprint Identification*, Advances in Computer Vision and Pattern Recognition, https://doi.org/10.1007/978-3-319-67681-4_4

Fig. 4.1 Contactless 3D
fingerprint acquisition using
coloured LEDs, **a** block
diagram, **b** sequential
positioning of LEDs. All six
LEDs are symmetrically
distributed along the circle

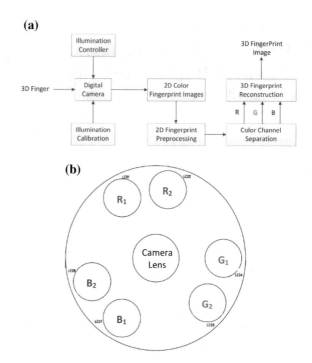

When one red, one green and one blue LEDs are simultaneously illuminated, one imaging shot is automatically acquired using the same program/software which controls the switching of LEDs. The imaging setup is calibrated offline and an iron ball was used to automatically locate and calibrate the LEDs positions using the method in [3]. The calibrated pixel positions (discussed in Chap. 3) are also made publicly available [4] for further research.

Three imaging shots are automatically acquired in sequence. The first two images are used for 3D fingerprint reconstruction and the last shot is the repetition of the first illumination position, which is used to *detect* the finger motion. The 1400 × 900 pixels' region of interest (ROI) from the acquired fingerprint is automatically segmented (Fig. 4.2), using similar approach as discussed in the last chapter. Each of the segmented contactless 2D fingerprint images is first subjected to preprocessing operations, including the removal of specular reflection, and these are discussed in the next chapter. Figure 4.1 illustrates simplified block diagram for coloured LEDs based on fingerprint image acquisition setup.

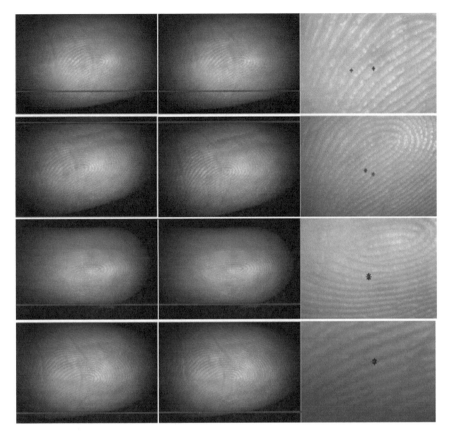

Fig. 4.2 Positional differences during 3D fingerprint imaging. The images in the first two rows are acquired image sample *with* motion. The images in the last two rows are sample without motion

4.2 Finger Motion Detection and Image Acquisition

Two imaging shots are acquired from the presented fingers to reconstruct the 3D fingerprint. Despite high-speed imaging, it is still possible to observe involuntary motion of fingers during successive imaging. In order to address such limitation, it is important to incorporate finger motion detection and accelerate image acquisition process. Two image shots are acquired in a short time interval of about 250 ms. In order to detect finger motion, third image shot is also acquired from presented finger, i.e., three consecutive imaging shots are acquired in an interval of about 800 ms. The third image shot is essentially the *repetition* of the image with the same illumination positions as the *first one* and is used to ascertain the finger motion. By computing the mean squared error (MSE) and key point positional differences in the first and the third images, fingerprint image samples with motion are ignored with thresholds (MSE > 5 and ΔX, $\Delta Y > 10$ pixels). Figure 4.2 illustrates such finger motion detection. The *sharpness* of each fingerprint image ROI is measured by computing the magnitude of image gradient. If the average magnitude of such fingerprint gradient image is larger than predetermined threshold, the acquired image can be considered as the blurred image and automatically discarded.

In addition to the finger motion, the specular reflection from finger surface also degrades the quality of reconstructed 3D fingerprint image. Reference [5] describes the use of SUV colour space to remove such specular reflection. This approach is quite effective in separating the specular and diffuse components into S channel and UV channels. However, this approach is computationally complex, i.e., required 0.322 s in our implementation, and is less attractive for online usage. We instead preferred a simplified approach which uses a predefined threshold to identify specular reflection pixels and replace or fill these identified pixels with average value of neighbourhood pixels.

This imaging approach is significantly faster than those using the setup in Fig. 4.1 of Chap. 3, and required 0.064 s in our implementation. The region of interest (ROI) images are automatically segmented from the acquired images using background detection. The ROI images from two colour images are split into RGB channels and such an image sample is shown in Fig. 4.3. Automatically generated six grey-level images are used to reconstruct 3D fingerprint as discussed in the next section.

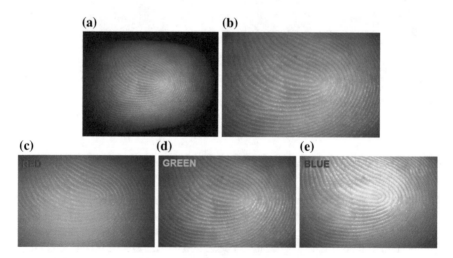

Fig. 4.3 **a** Acquired image sample from the setup in Fig. 4.1, **b** segmented colour ROI image, and (**c–e**), respectively, are the red, green and blue colour components of image sample in (**b**)

4.3   Reconstructing 3D Fingerprint Using RGB Channels

The colour (RGB) photometric stereo approach [10] is incorporated to reconstruct the 3D fingerprint. This approach works well under the assumption that finger surfaces are nearly Lambertian. Therefore, we consider Lambertian model for the surface reflectance where ρ represents its surface albedo. Let $I(x, y)$ represent the acquired 2D fingerprint images and $\mathbf{n}(x, y, z)$ be the *unit* surface normal vectors at respective finger surface. The LED illumination $\mathbf{L}(x, y, z)$ sources are calibrated as discussed in Chap. 3.

$$\mathbf{I} = \rho \mathbf{n} \cdot \mathbf{L} \tag{4.1}$$

where $\mathbf{I} = [I_1, I_2, \ldots, I_m]$, $\mathbf{L} = [L_1, L_2, \ldots, L_m]$ and m is numbers of LED light source. The surface normal vectors $\tilde{n} = \rho \cdot \mathbf{n}$ can be estimated from the following equation:

$$\tilde{\mathbf{n}} = \left(\mathbf{L}^{\mathsf{T}}\mathbf{L}\right)^{-1}\mathbf{L}^{\mathsf{T}}\mathbf{I} \tag{4.2}$$

The reflectance representing the surface albedo can be computed from the norm, i.e., $\rho = |\tilde{\mathbf{n}}|$. For the different RGB components of the acquired image $\mathbf{I}$, i.e., $\mathbf{I}_R$, $\mathbf{I}_G$ and $\mathbf{I}_B$, we can rewrite Eq. (4.1) as follows:

$$\mathbf{I}_R = \rho_R \mathbf{L}\mathbf{n} \tag{4.4}$$

$$\mathbf{I}_G = \rho_G \mathbf{L}\mathbf{n} \tag{4.5}$$

$$\mathbf{I}_B = \rho_B \mathbf{L}\mathbf{n} \tag{4.6}$$

The least squared solution from (4.2) can be used to recover surface normal for the above equations. The depth map of 3D fingerprints can be computed from the

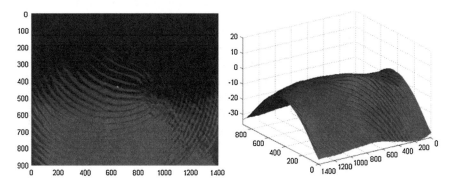

Fig. 4.4 Sample 3D fingerprint image acquired using colour photometric stereo with Frankot–Chellappa algorithm

integration of resulting gradients as discussed in Sect. 3.1.4 in the previous chapter. Figure 4.4 illustrates a sample 3D fingerprint image reconstructed from coloured photometric stereo using Frankot–Chellappa algorithm.

4.4   Reconstruction Accuracy and System Complexity

As underlined in the last chapter, it is very difficult to evaluate the precision of reconstructed 3D fingerprints as such evaluation requires measuring 3D ground truth from the presented 3D fingers, which is a very difficult task. In order to address such difficulty, we can attempt to evaluate the accuracy of recovered fingerprint surface normal, by computing the intensity error, as in [6]. Such error represents geometric accuracy for the synthetic 3D model and is similar to the 3D phantom targets like phantom eye [7] or 3D Ronchi targets, used to verify the resolution, distortion, magnification or spatial frequency response from a 3D imaging system. The mean squared error (MSE) is used to measure the average of the squared, between the ground truth normal and reconstructed 3D fingerprint surface normal. Any colour image can be separated into its RGB channel with the help of predetermined weights. Given a colour image F, it can be decomposed in three channels as follows:

$$F = \omega_1 * F_R + \omega_2 * F_G + \omega_3 * F_B \tag{4.7}$$

We evaluate the accuracy of surface normal by incorporating different weights to separate RGB channel images and compute the time complexity. One group of the sample was used [2] to perform experiments. When $\omega_1 = 0.3$, $\omega_2 = 0.4$ and $\omega_3 = 0.3$, the mean squared error value was observed to be minimum with MSE of 0.1087. This accuracy was also computed on the entire dataset [4]. Our results achieve MSE_{min} of 0.0696, MSE_{max} of 0.1615 and MSE_{avg} of 0.1202.

A synthetic model with known *depth* data, with some similarity to the model employed in [8], was developed and is shown in Fig. 4.5a. This model was employed to evaluate the accuracy of recovering 3D depth from the model. We also experimented with three different solutions to reconstruct 3D fingerprints while minimizing the error between the integrable and non-integrable gradient fields; Frankot–Chellappa with Fourier basis function and Poisson Solver using discrete cosine function were discussed in Sect. 3.1.4 in the previous chapter. In addition, the approach in [9] with shapelet basis was also evaluated. Computational complexity is also computed on both synthetic data and the ground truth (with fixed fingerprint image resolution of 1400×900). Our comparative results in Fig. 4.5 and Table 4.1 suggest that recovering 3D fingerprint using shapelets correlated approach [9] requires more computational time that can offer superior accuracy.

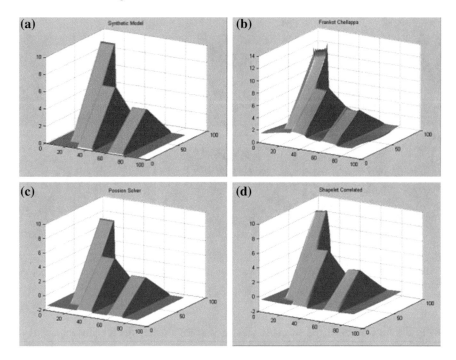

Fig. 4.5 The synthetic 3D model (**a**) employed to evaluate the reconstruction accuracy. The reconstructed 3D model using the Frankot–Chellappa method (**b**), using Poisson solver (**c**) and using shapelets (**d**)

Table 4.1 Comparative MSE and time complexity of different methods

Method	Frankot–Chellappa	Poisson solver	Shapelet correlated
Mean squared error	2.7749	1.9276	0.3061
O (synthetic data)	0.00869 s	0.02506 s	0.04627 s
O (ground truth)	0.10121 s	0.31368 s	1.27850 s

4.5 Influence from Finger Skin Contamination

Contactless 3D fingerprint imaging systems may have to cope with the contaminations on finger skin surfaces. Therefore, this section presents results from a brief study on the influence of coloured markings on finger skin on 3D images acquired using coloured photometric stereo. Six different fingerprint samples with and without red colour markings were acquired from the two different subjects/fingers, and these images are shown in Fig. 4.6. The corresponding reconstructed 3D fingerprint surfaces are illustrated in Fig. 4.7. As can be observed from the figure, that such red markings have very little or negligible effect on the reconstructed 3D fingerprint surface. Our further results in *matching* these contactless 3D fingerprints in Fig. 4.7,

(a)   **(b)**   **(c)**

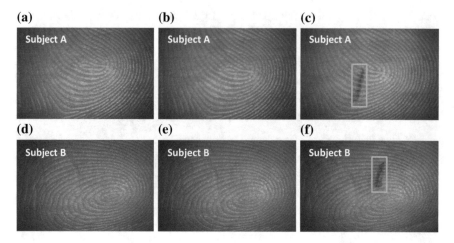

(d)   **(e)**   **(f)**

Fig. 4.6 a, **b,d**, **e** Two sample images from two different 2D fingerprints without red marking; (**c**, **f**) illustrate respective two different fingerprint samples, with the red marking on fingers, for the experiments

using the algorithm in [2], suggest that such colour markings are not expected to degrade the performance for the 3D fingerprint reconstruction, minutiae extraction and the 3D fingerprint matching.

It is useful to ascertain the influence of the same rakings/contaminations for the conventional contact-based fingerprint identification. Therefore, we also acquired the corresponding contact-based 2D fingerprints with and without markings from the same two subjects using URU 4000 fingerprint reader. As observed from the sample images in Fig. 4.8, the red colour marking can have some effects on the ridge–valley patterns observed from the contact-based 2D fingerprints. The matching score for sample (a1) and (a2) is **0.7050**, and for sample (b1) and (b2) is 0.6688. These scores can be compared with respective scores of 0.7603 and 0.7327 for matching respective fingerprints without markings. Our observations indicated degradation of genuine match scores while matching same fingers, with and without red colour markings on the finger skin, using contact-based fingerprint sensor.

4.6   Summary

This chapter discussed on the development of contactless 3D fingerprint imaging system using coloured photometric stereo. This approach was introduced to reduce complexity and address challenges associated with 3D fingerprint imaging due to involuntary finger movement. Another attractive single fixed camera-based approach for the acquisition of 3D fingerprints is to use elastomeric sensors, and this approach [7] can also address problems associated with involuntary finger motion. However,

Fig. 4.7 a 3D fingerprint using Frankot–Chellappa algorithm, **b** 3D fingerprint using Poisson solver, **c** 3D Fingerprint using shapelets correlated method. First two rows illustrate reconstructed fingerprints from images in Fig. 4.6a, b without red marking. The third row illustrates reconstructed fingerprint from the image in Fig. 4.6c with the red markings. Next two rows illustrate reconstructed fingerprints from images in Fig. 4.6d, e without red marking. Last row illustrates reconstructed fingerprint from the image in Fig. 4.6f with the red markings

Subject A   **Subject B**
(a1)   **(a2)**   **(b1)**   **(b2)**

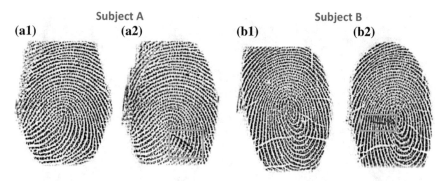

Fig. 4.8 (a1) and (a2) contact-based 2D fingerprint samples without marking and respective (b1)–(b2) contact-based 2D fingerprint samples with red markings

such an approach cannot benefit from the advantages associated with contactless 3D fingerprint imaging as it requires physical contact with elastomeric sensor. Acquired 3D fingerprint images, either using photometric stereo or any other methods discussed in Chap. 2, requires a range of postprocessing operations to recover the features. These details on preprocessing and postprocessing operations are discussed in the next chapter.

References

1. Spence A, Chantier M (2006) Optimal illumination for three image photometric stereo using sensitivity analysis. IET Vis Image Sig Proc 153(2):149–159
2. Lin C, Kumar A (2017) Tetrahedron based fast 3D fingerprint identification using colored LEDs illumination. In: IEEE transaction pattern analysis and machine intelligence, pp 1–10
3. Zhou W, Kambhamettu C (2002) Estimation of illuminant direction and intensity of multiple light sources. In: Proceedings of ECCV 2002. Springer, pp 206–220
4. The Hong Kong Polytechnic University 3D fingerprint Database V2, http://www.comp.polyu. edu.hk/~csajaykr/3Dfingerv2.htm, February 2018
5. Mallick SP, Zickler T, Belhumeur PN, Kriegman DJ (2006) Specularity removal in images and videos: a pde approach. In: Proceedings of ECCV 2006. Springer, pp 550–563
6. Xiong Y, Chakrabarti A, Basri R, Gortler SJ, Jacobs DW, Zickler T (2015) From shading to local shape. IEEE Trans Pattern Anal Mach Intell 37:67–79
7. Johnson MK, Cole F, Raj A, Adelson EH (2011) Microgeometry capture using an elastomeric sensor. ACM Trans Graphics (TOG) 30(4):46
8. Agrawal A, Raskar R, Chellappa R (2006) What is the range of surface reconstructions from a gradient field?. In: Proceedings of ECCV 2006. Springer, pp 578–591
9. Kovesi P (2005) Shapelets correlated with surface normals produce surfaces. Proc ICCV 2005:994–1001
10. Christensen PH, Shapiro LG (1994) Three dimensional shape from color photometric stereo. Int J Comput Vision 13(2):213–227

Chapter 5
3D Fingerprint Image Preprocessing and Enhancement

The raw 3D fingerprint data is generally accompanied by noise due to imaging artefacts or the limitations of 3D reconstruction algorithms. Therefore, preprocessing steps are required to suppress such undesirable influences of noise in the feature extraction process. There are many choices for representing 3D fingerprint data and are summarized in the next section.

5.1 3D Fingerprint Data Format and Representation

Contactless 3D Fingerprint image data can be represented in various forms. These are often dependent on the nature of sensors employed to recover 3D fingerprint data and the nature of intended applications. These representations can be classified into unorganized (cloud point, range image) and organized (polygon mesh, voxelized cloud) as discussed in the following.

- **Point Cloud:** The cloud point representation of 3D fingerprint data/image is simply a collection of 3D coordinates, i.e. representation 3D fingerprint surface using a set of vertices. The cloud point data representation is the most popular form for object representation in 3D computer vision applications. The cloud point representation cannot be used to discriminate between the inner and outer surfaces, e.g. dermis and epidermis, of 3D fingerprints. The 3D fingerprint raw data acquired using photometric stereo (previous chapter) and structured lighting approach is often represented in cloud point form.
- **Range Image**: Contactless 3D fingerprint image represented in range data form is essentially the distance between the 3D finger surface point from the image sensor. Therefore, the range data is a regular grid of 3D coordinates and does not have any specific topology.
- **Voxelized Cloud**: It can represent 3D fingerprint surface with a regular grid of volumetric pixels or voxels. A voxel therefore represents a value on a regular

© Springer Nature Switzerland AG 2018

A. Kumar, *Contactless 3D Fingerprint Identification*, Advances in Computer Vision and Pattern Recognition, https://doi.org/10.1007/978-3-319-67681-4_5

3D grid or voxelized cloud and its 3D coordinates are implicitly defined from its position on the grid. Voxelization is a useful representation for efficient 3D data processing. Voxelized cloud is often the preferred representation of the 3D fingerprint data acquired using the OCT.

- **Polygon Mesh**: The polygon mesh or triangle mesh is a topological representation of 3D finger surface data using a collection of 3D surface vertices and their connections. This representation provides orientation of vertices from the directions of their connections and is useful for rendering. The raw point cloud data is often converted to polygon mesh and is most widely used 3D data representation format in computer graphics.

Contactless 3D fingerprint raw data can enable (a) *full 3D view* fingerprint which provides 360° view 3D fingerprint data covering the dorsal/nail view, or (b) *2,5 D view* fingerprint which provides a 3D view of the regions visible from the sensor position. The premise of fingerprint biometric is based on the singularity of continuous friction ridges, which are not visible in the dorsal region. Therefore, unless explicitly stated, the contactless 3D fingerprint data essentially represents 2.5 D view fingerprint.

5.2 Contactless 3D Fingerprint Image Enhancement

Raw 3D fingerprint data is generally subjected to preprocessing operations to suppress the accompanying noise and enhance the clarity of ridge-valley features. The 3D fingerprint surface data in a point cloud form represents the height value (z) on the 2D plane (x, y). This cloud point data $z(x, y)$ is first subjected to the smoothing process, which first applies a median filter to suppress speckle-like noise. The resulting image is then subjected to a 3D variant of Laplacian smoothing which has been found to be quite effective in denoising point cloud data in many other applications [1]. For a vertex z_i within the 3D fingerprint image $z = f(x, y)$, with its neighbours z_j, the updated or new vertex values $\bar{z}_i$ are computed as follows:

$$\bar{z}_i = (1 - \epsilon)z_i + \frac{\epsilon}{\sum_j w_{ij}} \sum_j w_{ij} z_j \tag{5.1}$$

where w_{ij} is a finite support weighting function and is chosen as the inverse of the distance between the vertex z_i and its neighbours z_j, i.e. $w_{ij} = \left\| z_j - z_i \right\|^{-1}$. The reconstructed 3D fingerprint surface data, for the images shown in Chap. 3, were smoothed after 40 iterations with $\epsilon = 0.5$ and the neighbours j were chosen within ± 2 pixel in the x and y directions from vertex z_i.

The normal vectors of the cloud point 3D fingerprint data for the smoothed surface (above operations) are then computed by the gradient of $z = f(x, y)$. The surface normal vector is an upward normal with $(-g_x, -g_y, 1)$, where g_x and g_y are the gradients along x and y directions. These normalized surface normals are then used for the feature extraction process. Many effective feature extraction strategies, including the recovery of 3D minutiae templates, rely on the estimation of 3D surface curvature. Therefore, the estimation of surface curvature, for the 3D fingerprint data acquired from many popular 3D sensing techniques, is discussed in the next section.

5.3   Estimating 3D Fingerprint Surface Curvature

The differential geometric properties of 3D surface are invariant under rigid transformation and can be potentially used to compute intrinsic 3D *fingerprint surface* descriptors and can describe the nature of ridge-valley information in 3D space. These local differential 3D surface geometric properties, i.e. principal curvature, unit normal vector and surface directions, can themselves be used to match 3D fingerprints and such results are also discussed in Chap. 7. The principal surface curvature typically measures *local* bending of 3D fingerprint surface at each of the surface points while the principal surface directions indicate the directions of minimum and maximum 3D surface bending. Several algorithms are available in the literature to estimate the surface curvature using *local* surface fitting. Goldfeather and Interrante [2] have described three approaches, i.e. quadratic surface approximation, normal surface approximation and adjacent-normal cubic order surface approximation. Among these three approaches, adjacent-normal cubic order approximation algorithm has shown to be quite effective [3] for the contactless 3D fingerprint data and merits detailed discussion.

Let a given 3D fingerprint surface point be s with its normal N and its u neighbouring points be t_i with their normal vectors K_i where $i = 1, 2, \ldots u$. In the coordinate system with s as the origin $(0, 0, 0)$ and N as the z-axis, the position of neighbours t_i is (x_i, y_i, z_i) and the position of K_i is (a_i, b_i, c_i). Using the adjacent-normal cubic order algorithm, we attempt to automatically locate a surface that can fit the vertex and its neighbouring points such that

$$z = f(x, y) = \frac{a}{2}x^2 + bxy + \frac{c}{2}y^2 + dx^3 + ex^2y + fxy^2 + gy^3 \qquad (5.2)$$

The normal vector of the surface point s in the approximated surface can be written as follows:

$$N(x, y) = \left(f_x(x, y), f_y(x, y), -1\right) \qquad (5.3)$$
$$N(x, y) = \left(ax + by + 3dx^2 + exy + fy^2, bx + cy + ex^2 + 2fxy + 3gy^2, -1\right) \qquad (5.4)$$

The cubic order surface fitting, for both the neighbouring surface points and their normal, generates the following three equations for each of the surface points.

$$\begin{pmatrix} \frac{1}{2}x_i^2 & x_i y_i & \frac{1}{2}y_i^2 & x_i^3 & x_i^2 y_i & x_i y_i^2 & y_i^3 \\ x_i & y_i & 0 & 3x_i^2 & 2x_i y_I & y_i^2 & 0 \\ 0 & x_i & y_i & 0 & x_i^2 & 2x_i y_i & 3y_i^2 \end{pmatrix} \Omega = \begin{pmatrix} z_i \\ -\frac{a_i}{c_i} \\ -\frac{b_i}{c_I} \end{pmatrix} \tag{5.5}$$

where $\Omega = [a\,b\,c\,d\,e\,f\,g]^T$ is the coefficient vector of the cubic surface. Above equation is an overdetermined equation system and can be written in the following form:

$$K\Omega = R \tag{5.6}$$

where K is $3u \times 7$ matrix (from left-hand side of Eq. 5.3) and R is $3u \times 1$ vector. We can apply least square fit to find the best solution for Eq. (5.6) and construct Weingarten curvature matrix W for the fitted surface using only three coefficients.

$$W = \begin{pmatrix} a & b \\ b & c \end{pmatrix} \tag{5.7}$$

The eigenvalues of Weingarten matrix are the maximum and minimum principal curvature of the surface (k_{max} and k_{min}), and their eigenvectors are the principal direction vectors (h_{max} and h_{min}) which can be directly computed. The shape index of a surface at vertex s can quantify the local 3D shape of the 3D fingerprint surface. One effective approach to quantify the 3D curvature information is to compute the shape index $C_i(s)$ values [4], which is independent of scale and can be estimated as follows.

$$C_i(s) = \frac{1}{2} - \left(\frac{1}{\pi}\right) atan\left(\frac{k_{max} + k_{min}}{k_{max} - k_{min}}\right) \tag{5.8}$$

The surface curvature map generated from the quantification of local surface index is in the interval [0–1] and can also be directly employed to match two 3D fingerprint surfaces which is discussed in Chap. 7. Figure 5.1 illustrates a sample grayscale 2D image representing surface curvature map for a 3D fingerprint.

5.4 Contactless Fingerprint Image Preprocessing

Many contactless 3D fingerprint imaging techniques are capable of generating 2D fingerprint images from the data used to reconstruct 3D fingerprints, e.g. using stereo imaging or photometric stereo approach. Simultaneous acquisition of contactless 2D fingerprints can be ensured with most of other 3D imaging techniques as the additional cost for such imaging sensor is relatively very small. Such simultaneously generated or acquired 2D contactless fingerprint images can be used to improve the

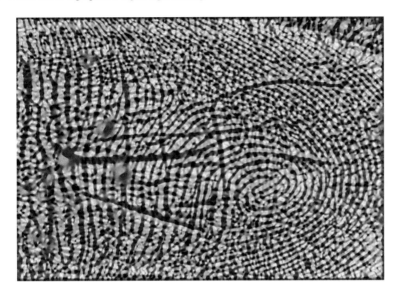

Fig. 5.1 Surface curvature map represented as 2D image from a sample 3D fingerprint cloud point data

Fig. 5.2 Contrast enhancement using high-pass homomorphic filtering

matching accuracy for 3D fingerprint identification systems. However, such contactless 2D fingerprint images generally have low contrast and need specialized methods to improve the contrast, before these images can be subjected to conventional fingerprint enhancement algorithms for contact-based fingerprints (e.g. using Gabor filters [5]) and used to generate match scores using such conventional fingerprint matching algorithms. Therefore, image contrast enhancement for such contactless fingerprint images is essential and can be achieved using homomorphic filtering as discussed in the following.

The image formation model for contactless fingerprint $I(x, y)$ is based on the surface reflectance and incident illumination as discussed in Chap. 3, $I(x,y) = R(p(x, y), q(x, y)) * I_0(x, y)$. where $0 \leq I_0(x, y) < \infty$ and $0 \leq R(p(x, y), q(x, y)) \leq 1$. Figure 5.2 illustrates main steps for the contactless fingerprint contrast enhancement using homorphic filtering using a high-pass filter. The key objective of homomorphic filtering [6] is to separate these two components using a logarithmic transform. These two transformed components can then be subjected to a high-pass filter $G(u, v)$ to enhance the reflectance part while reducing contributions from the illumination component. Generation of contrast-enhanced 2D fingerprint image $I_e(x, y)$ using such high-pass homomorphic filtering can be summarized using the following equation:

$$I_e(x, y) = \exp\left[\mathcal{F}^{-1}[G(u, v) \cdot \mathcal{F}[\ln[I(x, y)]]]\right] \tag{5.9}$$

$$G(u, v) = \left(\frac{1}{1 + \left(\frac{D_0}{D(u,v)}\right)^{2n}}\right) \tag{5.10}$$

where $\mathcal{F}$ is the Fourier transform operator, $G(u, v)$ is the high-pass filter with D_0 cut-off frequency and n represents the order of this filter. The term $D(u,v)$ represents the Euclidean distance from the origin in FFT, i.e. $sqrt(x.^2+y.^2)$. The filter $G(u, v)$ can also be designed to *suppress* the contributions from the low-frequency part representing the illumination component $I_0(x, y)$. This approach is more effective in the presence of strong external illumination like those acquired under the illumination setup for the photometric stereo. Therefore, a bandpass filter $H(u, v)$, resulting from the combination of low-pass and high-pass filter, $G_1(u, v)$ and $G_s(u, v)$ as shown in Fig. 5.3, is more appropriate choice for the contrast enhancement.

$$H(u, v) = \min\left(\frac{1}{1 + \left(\frac{D_0}{D(u,v)}\right)^{2n}}, \frac{1}{1 + \left(\frac{D(u,v)}{D_1}\right)^2}\right) \tag{5.11}$$

where D_0, D_1 and *n* are the high-pass cut-off frequency, low-pass cut-off frequency and order of filter. The enhanced fingerprint image is generated from the exponent of the inverse Fourier transform of the frequency domain image as shown in (5.9).

In order to achieve normalized greyscale values in 0–1 or 0–255 range, the enhanced image $I_e(x, y)$ is also subjected to adaptive histogram equalization. Contactless 2D fingerprint images, which are simultaneously generated or acquired with 3D fingerprint images, are preprocessed using homomorphic operation before subjecting to the conventional fingerprint enhancement [5, 7] steps. The surface curvature images from the 3D fingerprint preprocessing operations do not have low-contrast concerns in the low-frequency part and more frequent high-frequency noise is eliminated using the preprocessing steps discussed in Sect. 5.3 of this chapter.

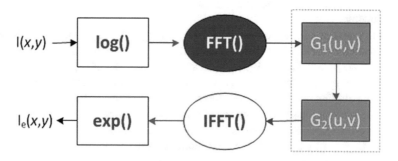

Fig. 5.3 Contrast enhancement using high-pass homomorphic filtering

The next step on the preprocessed 3D fingerprint image is to generate templates that can represent unique features. In the context of 2D fingerprint images, the minutiae features are most popular and widely used in the law enforcement around the world. The representation, recovery and matching of these features in 3D spaces are discussed in the next chapter.

References

1. Belyaev A (2006) "Mesh smoothing and enhancing. curvature estimation," Saarbrucken. www.mpi-inf.mpg.de/~ag4-gm/handouts/06gm_surf3.pdf
2. Goldfeather J, Interrante V (2004) A novel cubic-order algorithm for approximating principal direction vectors. ACM Trans Graphics 23(1):45–63
3. Kumar A, Kwong C (2013) Towards contactless, low-cost and accurate 3D fingerprint identification. In: Proceedings of CVPR 2013, Portland, USA, pp 3438–3443
4. Dorai C, Jain AK (1997) COSMOS—a representation scheme for 3D free-form objects. T-PAMI, pp 1115–1130
5. Chen Y (2009) Extended feature set and touchless imaging for fingerprint matching, Ph.D. thesis, Michigan State University
6. Gonzalez RC, Woods RE (2018) Digital image processing, 4th edn. Pearson
7. O'Gorman L, Nickerson JV (1989) An approach to fingerprint filter design. Pattern Recogn 22:29–38

Chapter 6
Representation, Recovery and Matching of 3D Minutiae Template

Contactless 3D fingerprint images can reveal a variety of depth or shape-related features and some of these have been investigated in the literature. The features extracted from the 3D fingerprint images can also be limited by the nature of sensing technique and the image resolution. The acquisition of 3D fingerprint image in [1] uses multiple views of a fingerprint from different viewpoints and under different illuminations. Such shape from silhouette method to reconstruct contactless 3D fingerprint has not been successful in precisely recovering the 3D fingerprint ridge details and therefore not investigated for 3D fingerprint matching. The 3D fingerprint images acquired from a range of sensing techniques, e.g. structured lighting or photometric stereo, can provide 3D fingerprint ridge–valley depth details. Recovery and matching of only fingerprint depth or height, as attempted in [2] on database of 11 subjects, is not expected to offer accurate results or recover most discriminant information, especially while matching 3D fingerprints from large lumber of subjects.

Among the variety of popular features, e.g. minutiae, ridge feature map, orientation field, etc., available for the fingerprint recognition, minutiae features are considered as most reliable [3] and also widely employed by the law enforcement experts and most of the commercial fingerprint systems available today. Their location in x–y plane, type and orientation is considered to be the most distinctive for conventional 2D fingerprint recognition. Therefore, it is imperative that we also attempt to further improve such distinctiveness these minutiae features by locating and matching them in 3D spaces. This can by incorporating two additional features for the representation of minutiae in 3D spaces: (a) minutiae height z and (b) its 3D orientation ϕ. This chapter systematically details the recovery and composite matching strategy for resulting 3D minutiae template representations. The approach detailed [4, 5] for 3D minutiae representation and matching in this chapter is generalized with wide applications, and can also be employed to match 3D fingerprint images reconstructed from many methods, including those from the structured lighting imaging, photometric stereo or OCT.

© Springer Nature Switzerland AG 2018

A. Kumar, *Contactless 3D Fingerprint Identification*, Advances in Computer Vision and Pattern Recognition, https://doi.org/10.1007/978-3-319-67681-4_6

6.1   Conventional 2D Fingerprint Minutiae Representation

The minutiae features from the 2D fingerprints are widely used and it will be useful to review on their representation and matching in 2D space. Another reason for this discussion is that the simultaneously acquired of available contactless fingerprint images can also be used to recover such features and further improve the performance for 3D fingerprint identification.

Conventional methods of 2D fingerprint image preprocess first employ enhancement [6], ridge detection and detection of minutiae features from the discontinuities in the ridge patterns as detailed in the literature [3]. These minutiae features form the feature representation or the fingerprint template. Two such arbitrary 2D fingerprint templates, say P and Q, can be matched to generate a match score as follows. We first select a minutiae pair, consisting of a minutia from the reference template (Q) and a minutia from the query template (P), to generate match distances between them using the alignment-based approach as illustrated in Figs. 6.1 and 6.2. All the *other* minutiae in template P are also converted to the spherical coordinate as $[r, A_s, A_\theta, T]$, with reference to the centre at the chosen reference minutia in Q and aligned the angle with θ_r.

$$r = \sqrt{(x - x_r)^2 + (y - y_r)^2} \tag{6.1}$$

$$A_s = \operatorname{atan2}\left(\frac{y - y_r}{x - x_r}\right) - \theta_r \tag{6.2}$$

$$A_\theta = (\theta - \theta_r) \tag{6.3}$$

where r is the distance of respective minutiae with the reference minutia, A_s is the angular separation of the minutia, A_θ is the new orientation of the minutia with respect to the chosen reference minutia and T is the type of minutia. If the difference

Fig. 6.1 a Representation of a sample minutia (x, y, θ) in 2D space with reference to the origin and representation of second minutia (x', y', θ') with reference to same reference or origin in 2D space. Two minutiae of *same* type (T) are considered as matched when their all relative distances $(r, A_s, A_\theta,)$ are smaller than some threshold (threshold 1, threshold 2, threshold 3)

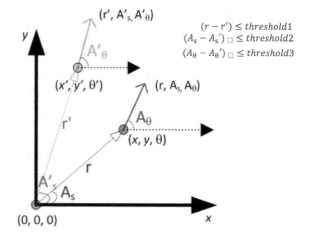

Fig. 6.2 Relative representation of a sample minutiae (x_2, y_2, θ_2) in 2D space with reference to another minutiae (x_1, y_1, θ_1) using relative measurements r, A_s, A_θ

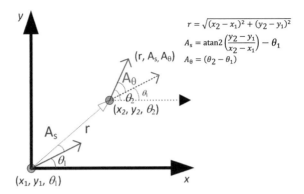

between $[r_i, A_{si}, A_{\theta i}, T_i]$ in P and $[r_j, A_{sj}, A_{\theta j}, T_j]$ in Q is smaller than a predetermined threshold and $T_i = T_j$, then the minutia i in P and minutia j in Q are considered as matched pair. The matching score is generated by popular approach as follows:

$$S_{2D} = \frac{m^2}{M_P M_Q} \tag{6.4}$$

where m is the total number of matched minutiae pairs with a chosen reference minutia from template Q and M_P, and M_Q is the number of minutiae in query and template image, respectively. The maximum of all the possible match scores (6.4), generated by selecting each of the available minutiae as the reference minutia, is selected as the final matching score between fingerprint image P and Q.

6.2 Minutiae Representation in 3D Space

Conventional fingerprint templates typically include 2D minutiae details (x, y, θ, q) consisting of position of the minutiae (x, y) along with the angle θ representing the orientation of the minutiae and the quality q of minutiae [7]. We can enrich such 2D minutiae templates to include 3D details by incorporating two additional features: minutia height z and elevation angle ϕ. The value z is the height of the vertex on the reconstructed 3D fingerprint surface at a position (x, y) while θ and ϕ represent the minutiae orientation in spherical coordinates with unit length 1. The angle ϕ can be computed by tracing the reconstructed 3D fingerprint surface at minutiae locations along the direction of θ, which is available from 2D minutiae details (x, y, θ) as illustrated in Fig. 6.3. The 3D surface curvature image, obtained from the 3D fingerprint cloud point data, is used to extract 3D fingerprint ridge structure (as explained in the next section). In case of bifurcation type of minutiae,

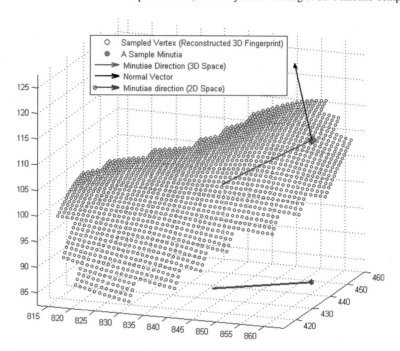

Fig. 6.3 Representation of minutia feature in 3D space on a reconstructed 3D fingerprint surface. The purple arrow on 2D or x–y plane illustrates the orientation of original 3D minutia in a conventional 2D fingerprint template. This 2D minutia orientation (purple arrow) is defined on a local ridge surface (blue dot) and can be used to estimate the 3D minutia orientation (red arrow)

such local ridge surface is utilized (or masked) for tracing 3D orientation. The angle ϕ is then computed by estimating the principle axes [8] of the masked ridge surface. However, for the end type minutiae, the local valley surface is masked since the direction θ is, in this case, is pointing in the outward direction. The extended minutiae representation in 3D space can be denoted/localized as (x, y, z, θ, ϕ), where (x, y, θ) is the respective minutiae representation in 2D space. In our further discussion, we refer to this minutiae representation (x, y, z, θ, ϕ) as the 3D minutia.

6.3 Recovering Minutiae in 3D Space from the 3D Fingerprint Images

Source 3D fingerprint image data can be used to generate local surface curvature images which can provide a 2D representation of 3D source information. Figure 6.4 illustrates such sample 3D curvature images using the shape index as detailed in Eq. (5.8) of the *previous* chapter. These images encode local 3D fingerprint ridge–valley

Fig. 6.4 Sample images representation of local 3D surface curvature using shape index images

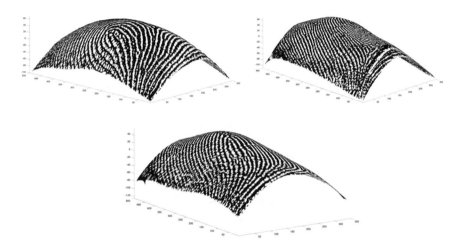

Fig. 6.5 Sample 3D fingerprint images illustrating ridge–valley structure identified from the local surface curvature values

information in 0–1 range and are shown as grey-level images. This information can be used to identify the 3D fingerprint ridges and the valleys. When a 3D surface is flat, the shape index values in (6.4) are expected to be zero. For the 3D fingerprint ridges, the shape index values are expected to be smaller than 0.5 (convex surface). Similarly, when the shape index values are higher than 0.5, the 3D surface is expected to be part of 3D fingerprint valleys.

Therefore, a simple the binarization of surface curvature images, i.e. If $C_i(s) \leq 0.5$ assign as 1 (ridge) else 0 (valley), can be used to label the 3D point cloud data as ridge–valley image.

- Shape Index for flat surface $\rightarrow$ 0
- Valley $\rightarrow$ If shape index > 0.5
- Ridge $\rightarrow$ If shape index ≤ 0.5.

Such sample 3D fingerprint images describing the ridge–valley details are shown in Fig. 6.5.

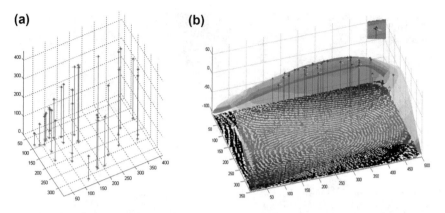

Fig. 6.6 Localization of minutiae in 3D space by incorporating minutiae height in (**a**) and illustration of 3D fingerprint with recovered minutiae locations in 3D space (**b**)

The binary images representing the ridge–valley structure of 3D fingerprints can also be used to detect the nature of 3D minutiae and their spatial locations. This step requires thinning of binarized ridge–valley images and usage of 3×3 mask to detect minutiae endings or bifurcations. Localization of minutiae in 2D space, along with its orientation θ in x–y plane, can be achieved by steps that are similar to those employed in conventional 2D fingerprint images, i.e. steps after that binarization of the enhanced D fingerprint images. Once the (x, y) locations and the type of minutiae are detected, the next step is to compute its height (z). This step is quite straightforward as the processed cloud point 3D fingerprint data already provides height for every (x, y) locations in 3D space (Fig. 6.6). Figure 6.6a illustrates such estimation of height for minutiae points and Fig. 6.6b illustrates localization of such minutiae, along with their orientation, in 3D space. Every minutiae at (x, y, z) locations and with θ orientations (in x–y plane) is fully localized in 3D space by computing the elevation angle ϕ. Computation of elevation angle (ϕ) of minutiae represented by a vector in 3D space, given its projection in 2D plane (θ) and the orientation of unit surface normal vector, is computed by tracing its orientation (Fig. 6.3) as discussed in the previous section.

6.4 Matching 3D Minutiae Templates

Each of the 3D minutiae template, consisting of (x, y, z, θ, ϕ) values for every detected minutia, is required to be robustly matched to establish the identity of 3D fingerprints. The key challenge in matching such minutiae template results from the lack of registration data, i.e. every sensor can measure the relative minutiae features with different references and the fingers presented for the contactless 3D sensors can have rotational and/or translation changes even among the successive 3D fingerprints from same subjects. Any robust 3D minutiae template matching algorithm should also consider the influence of missing or spurious minutiae, which are frequently observed from contactless 3D fingerprint images. An effective approach to generate match scores, which can also consider adverse influence from missing and/or spurious minutiae, between 3D minutiae templates is introduced [5] in the following.

Let us consider two arbitrary 3D minutiae fingerprint templates, say P and Q respectively with M_P and M_Q number of minutia, for generating a quantitative match score. We first select a (any) reference minutia from the template P and template Q; (a) All other minutiae in template P are then transformed to the spherical coordinates using respectively the chosen minutia from this template as the reference, i.e. we align all other minutiae in template P with the x-axes and z-axes (using θ and ϕ values of the chosen reference minutiae). Similar to the previous step (a), we transform all the other minutiae in Q to spherical coordinates and align with x-axes and z-axes of chosen reference minutia in this template. This alignment step ensures that the *reference* minutiae location (in both template P and Q) for the alignment can serve as a universal origin/reference (Fig. 6.7) to measure other/relative minutiae distances in the respective templates, *when* the same 3D minutia from the same finger appears in both templates, i.e. the case when the chosen reference minutia in P and Q are the same and part of genuine comparisons. If an aligned minutia is represented as $m_r = [x_r, y_r, z_r, \theta_r, \phi_r]$ in template P, the relative representation of other 3D minutiae in template P,[1] say m (see Figs. 6.7 and 6.8 to visualize relative representation of two 3D minutiae), can be denoted as $m = [r, A_s, A_\theta, A_g, A_\phi]$, where r is the radial distance with reference minutiae, A_θ is the azimuth angle and A_ϕ is the elevation angle that localizes the minutiae m in 3D plane, while A_s and A_g are the azimuth and the elevation angles respectively that localize the radial vector r (with respect to reference minutiae m_r) in 3D space. Let $R_z(\theta)$ and $R_y(\phi)$ be the rotation matrix along z and y direction in Cartesian coordinate, and sph(x, y, z) be the Cartesian to Spherical coordinate transformation with unit length 1:

$$R_z(\theta) = \begin{bmatrix} \cos\theta & -\sin\theta & 0 \\ \sin\theta & \cos\theta & 0 \\ 0 & 0 & 1 \end{bmatrix}, \quad R_y(\phi) = \begin{bmatrix} \cos\phi & 0 & -\sin\phi \\ 0 & 1 & 0 \\ \sin\phi & 0 & \cos\phi \end{bmatrix} \quad (6.5)$$

$$\text{sph}([xyz]) = \left[\text{atan2}(y, x)\ \sin^{-1} z\right] \quad (6.6)$$

where atan2 is the four-quadrant inverse tangent function [9]. The parameters for the relative representation (feature vector) of minutiae m are computed as follows:

$$r = \sqrt{(x - x_r)^2 + (y - y_r)^2 + (z - z_r)^2} \quad (6.7)$$

$$[x'y'z']^T = R_y(-\phi_r)R_z(-\theta_r)\frac{1}{r}[x - x_r y - y_r z - z_r]^T \quad (6.8)$$

$$[A_s A_g] = \text{sph}([x'y'z']) \quad (6.9)$$

---

[1] Also in template Q since the reference minutiae have been aligned to serve as the universal reference/origin.

$$[A_\theta A_\phi] = \mathrm{sph}\left(\left(R_y(-\phi_r)R_z(-\theta_r)\left(\mathrm{sph}^{-1}([\theta\phi])\right)^T\right)^T\right) \tag{6.10}$$

We perform the same process as discussed in above paragraph for the minutiae in template Q, i.e. using an arbitrarily selected minutia in Q, align all other minutiae in template Q and compute the relative representation $(r_{Qi}, A_{s\,Qi}, A_{\theta\,Qi}, A_{g\,Qi}, A_{\phi\,Qi})$ of all other minutiae using (6.5)–(6.10). Next step is to identify the number of matched minutiae, if any, between the aligned minutiae in P and Q. Two 3D minutiae in the two fingerprint template P and Q are considered as matched pair if the difference between

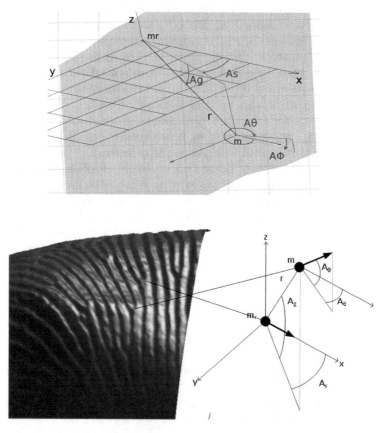

Fig. 6.7 Relative localization of a 3D minutiae with a reference minutiae (m_r) in a given 3D fingerprint template using the graphical illustration of relative distances/angles between the reference 3D minutia (m_r) and other 3D minutia (m). The x-axis of Cartesian coordinate is aligned with the direction of m_r (the bold arrow). The azimuthal angles A_s and A_θ are in the range of [0, 360] degree. The polar angles A_g and A_ϕ are in the range of [−90, 90] degree. It may be noted that the magnitude/value of $(r, A_s, A_\theta, A_g, A_\phi)$ are *exaggerated* simply to illustrate these values clearly (rather than use exact from left-hand size image) and most of the finger surface is convex. In the left figure, the green line for m illustrates the orientation of 3D minutiae while the red line illustrates its projection in x–y plane

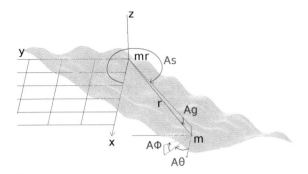

Fig. 6.8 Another sample 3D fingerprint minutiae template to illustrate relative localization of a 3D minutiae m, using feature vector $\mathbf{r}$ ($r, A_s, A_g, A_\theta, A_\varphi$), with the reference minutia m_r

their feature vectors ($r_{Pi}, A_{s\,Pi}, A_{\theta\,Pi}, A_{g\,Pi}, A_{\phi\,Pi}$) and ($r_{Qi}, A_{s\,Qi}, A_{\theta\,Qi}, A_{g\,Qi}, A_{\phi\,Qi}$) is smaller than a given threshold (or tolerance limit).

$$\Delta r = \left| r_{Pi} - r_{Qi} \right| \tag{6.11}$$

$$\Delta_{A_s} = \min\left(\left| A_{s\,Pi} - A_{s\,Qj} \right|, 360° - \left| A_{s\,Pi} - A_{s\,Qj} \right| \right) \tag{6.12}$$

$$\Delta_{A_\theta} = \min\left(\left| A_{\theta\,Pi} - A_{\theta\,Qj} \right|, 360° - \left| A_{\theta\,Pi} - A_{\theta\,Qj} \right| \right) \tag{6.13}$$

$$\Delta_{A_g} = \left| A_{g\,Pi} - A_{g\,Qj} \right| \tag{6.14}$$

$$\Delta_{A_\phi} = \left| A_{\phi\,Pi} - A_{\phi\,Qj} \right| \tag{6.15}$$

If $\Delta r \le \mathrm{th_r}$, $\Delta_{A_s} \le \mathrm{th}_{A_s}$, $\Delta_{A_\theta} \le \mathrm{th}_{A_\theta}$, $\Delta_{A_g} \le \mathrm{th}_{A_g}$ and $\Delta_{A_\phi} \le \mathrm{th}_{A_\phi}$, the minutiae pair from P and Q are considered as matched. Total numbers of such matched minutiae pairs are computed. The *maximum* number of matched minutiae pairs, among all the possibilities of every minutiae in P and Q being used as a reference, is used to compute the matching score. Figure 6.9 illustrates an example from such automatically selected reference. In other words, we compute all the possible combinations of selecting reference minutiae in template Q, for every chosen minutiae in P as reference, and generate the best possible match score as illustrated in Fig. 6.10. This step can ensure that adverse influence of missing or spurious minutiae is minimized during the computation of final match score. This (normalized) matching score between the two 3D minutiae templates is computed as follows:

$$S_{3D} = \frac{m^2}{M_P M_Q} \tag{6.16}$$

where m is the total number or the maximum number of *matched* minutiae pairs using (6.11)–(6.15).

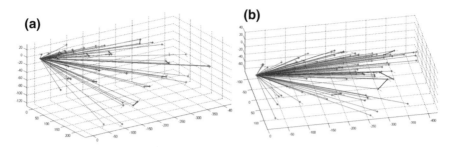

Fig. 6.9 Matching two genuine 3D minutia templates in (**a**) and the imposter templates in (**b**) from template P (in red colour) and template Q (in blue colour). The colour lines are links from the reference minutia while the black lines illustrate the minutiae pairs that are regarded as matched

6.4.1 3D Minutiae Quality

Quantification of minutia quality can represent the confidence in the accurate recovery of minutia from the presented finger images. In case of 2D fingerprint images, there are several measures to quantify the minutiae quality [7, 10] and these are largely related to the quality of greyscale images or ridge–valley regions which determine the accuracy in localization or recovery of minutiae features. In case of 3D fingerprints, the 3D minutiae also includes (x, y, θ) features as for 2D fingerprints and therefore the minutiae quality extracted from the surface curvature images (Fig. 6.5) can also reflect the quality of 3D minutiae and was investigated during this research. Quantification of 3D minutiae quality, independent to the method of 3D imaging, should also include some measure of confidence in the accurate recovery of height (z) and elevation angle (ϕ). Development of such unified 3D minutia quality representation yet to be addressed and part of ongoing research.

The quality of minutiae recovered from the 3D surface curvature images, extracted from 3D fingerprint data, can be incorporated to estimate minutia quality and the number. Reference [10] from NIST provides details on the reliability measure that represents minutiae and image quality in five different quality levels, i.e. from level 0 to 4 (worst to best). The template file from MINDCT [7] provides the minutiae quality number Q_m (0–99 range) and when image quality L is 4 (highest), the *minutiae* quality $Q_m \geq 50$, i.e. for $L = 4$, $Q_m = 0.50 + (0.49 * R)$, where R represents greyscale reliability and is computed from the variance and mean value greyscale pixels surrounding the minutiae point (more details in [10]). A study on minutiae quality using this measure, on a database of 135 different clients 2D and 3D fingerprints acquired using the setup detailed in Chap. 3, was performed to ascertain the number of minutiae or varying quality level from different preprocessed images. Table 6.1 summarizes the statics from the minutiae file where Q_{3DS} represents minutiae quality in 3D fingerprints generated from the 3D surface curvature images, Q_{2D7u} represents

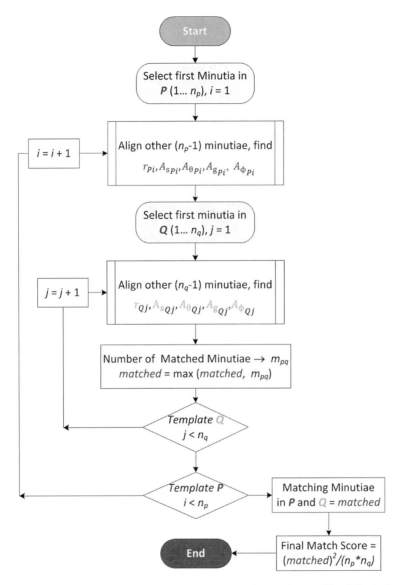

Fig. 6.10 Matching two 3D minutiae templates to generate the match score. The difficulty in synchronizing or aligning different minutiae among different capture can be addressed by considering every 3D minutiae in the given template/capture as a reference and using the reference 3D minutiae which generates the maximum number of matched minutiae (or match scores) as the final reference for computing the match score

minutiae quality from the unified image generated from a set of 7 2D images acquired for 3D reconstruction and Q_{2D7} represents minutiae quality from 7 images acquired

Table 6.1 Number of minutiae and minutiae quality distribution

	Q_{3DS} *for L=4*	Q_{3DS} *for* L $\geq$ 3	Q_{2D7u}	Q_{2D7}
Average number of minutiae per file	22	77	40	34
Variance (number of minutiae per file)	90	407	184	178
Average quality of minutiae	85	54	78	84
Variance (quality of minutiae)	32	441	75	79

for the 3D reconstruction. The numbers in this table are specific to the database but indicate that the quality of 2D fingerprint minutiae can be used for improving 3D fingerprint matching performance. One such possible approach is discussed in the following section.

6.4.2 3D Minutiae Selection

The limiting accuracy of preprocessing and 3D reconstruction algorithm, finger movement and sensor/surface noise, often influences the accuracy and reliability of reconstructed 3D minutiae. Therefore, this section introduces an effective approach which considers reliability recovered 3D minutiae during the template matching. In this approach, the recovered 3D minutiae which can be considered to have higher reliability are only utilized to achieve higher confidence during the template matching process.

This approach uses simultaneously acquired or available contactless 2D finger-print image to infer the quality of 3D minutiae. In the case of photometric stereo based 3D fingerprint imaging, multiple noisy 2D finger images acquired for 3D reconstruction can themselves be employed to infer the reliability of 3D minutiae recovered at any specific x–y location. If a 2D minutia from a spatial region is detected in more than one image (seven in our case/setup, Fig. 6.11), among the multiple 2D fingerprint images acquired with different LED illuminations for the respective 3D fingerprint reconstruction, then its *corresponding* 3D minutiae from this region can be considered as more reliable. Such an algorithm employed to automatically select 3D minutiae with higher reliability is discussed in the following.

Let the list of minutiae extracted from the 2D fingerprint image under the first LED illumination be $L=\{m_1, m_2, \ldots m_n\}$, where $m = [x, y, \theta, q]$. We first initialize the *counted list* $CL=\{cm_1, cm_2, \ldots cm_n\}$, where $cm=[x, y, \theta, q, c]$, c is the number of occurrences and is set to 1. For each minutiae m_i from the fingerprint image under the second to the seventh LED illumination, the CL is updated as follows: Let $\{TL\}$ be a subset of CL such that $x_{cm,k} - x_i, y_{cm,k} - y_i^2 \leq$

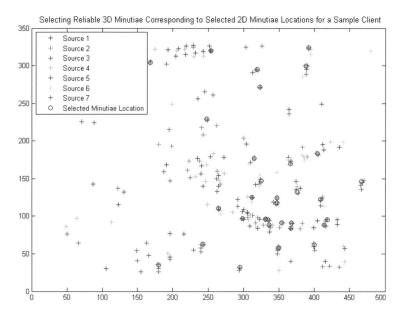

Fig. 6.11 The location of minutiae (endpoint or bifurcation), from a sample clients 2D fingerprint images under different illuminations, with lighting (cross) and the clustered location (circle)

$k1$ and $\min\left(\left|\theta_{\mathrm{cm,k}} - \theta_{\mathrm{i}}\right|, 360 - \left|\theta_{\mathrm{cm,k}} - \theta_{\mathrm{i}}\right|\right) \leq k2$ where $\mathrm{cm}_{\mathrm{k}} \in \mathrm{CL}$, then we update the cm_{t} such that $c_{\mathrm{t}} \geq c_{\mathrm{i}}$ for all $\mathrm{cm}_{\mathrm{i}} \in \{\mathrm{TL}\}$. x, y, θ, q value of updated cm_{t} will be the average value of existing cluster members and new member, and c will be increased by one. We choose $k1$ as 4 since the minutiae location in different (LED) fingerprint images would not shift too far away and the square root of 4 (=2) which is slightly smaller the half width of the observed ridge (~5 pixel in the employed fingerprint images). The constant $k2$ is set to 25 for the acceptable angle difference which can help to ensure that the clusters have similar direction while $k3$ is set to 32 to decrease the overlapping/double cluster with the similar direction and location. After updating CL, we picked the subset of CL as DL with $c \geq 2$. If two clusters groups are too close, we merge them together to reduce the possibility that a single minutia is recovered as two minutiae. The final list of minutia is the merged list of DL which is assumed (shown/suggested from the experiments) to have higher reliability and matching. The 3D minutiae corresponding to these 2D minutiae locations are employed during the matching stage. This strategy to select reliable 3D minutiae is summarized by the algorithm S3DM.

Algorithm: S3DM

Input: List of minutiae from 7 LED images $\{L_1,...,L_7\}$, where $L_i=\{m_1, m_2,... m_n\}$ and $m_k = [x_k, y_k, ,\vartheta_k, q_k]$

$CL := \{\}$ // the tuple of CL is $[x, y, \vartheta, q, c]$ where c is number of count

for each minutia m_k in L_1 **do**

       $CL := CL \cup [x_k, y_k, ,\vartheta_k, q_k, 1]$

end for

for each minutia m_k in $L_2, L_3,...,L_7$ **do**

     $TL := \{\}$

     for $m_{CL} \in CL$ **do**

         if$\|x_{CL} - x_k, y_{CL} - y_k\|^2 \leq k1$ and $\min(|\theta_{CL} - \theta_i|, 360 - |\theta_{CL} - \theta_k|) \leq k2$

           $TL := TL \cup m_{CL}$

         end if

     end for

     if $TL \neq \{\}$

         $maxC := 0$

         for $m_{TL} \in TL$ **do**

            if $c_{TL} > maxC$

               $m_t := m_{TL}$

               $maxC = c_{TL}$

            end if

         end for

         $[x_s, y_s, \theta_s, q_s] := (c_t*[x_t, y_t, \theta_t, q_t] + [x_k, y_k, ,\theta_k, q_k])/(c_t+1)$

         $m_t := [x_s, y_s, \theta_s, q_s, c_t+1]$

     else

         $CL = CL \cup [x_k, y_k, ,\vartheta_k, q_k, 1]$

     end if

end for

$DL := \{\}$

for $m_{CL} \in CL$ **do**

     if $c_{cl} >= 2$

         $DL := DL \cup m_{cl}$

     end if

end for

for $m_{DL} \in DL$ **do**

     for $m_k \in DL$ **do**

         if **if**$\|x_{DL} - x_k, y_{DL} - y_k\|^2 \leq k3$ and $\min(|\theta_{DL} - \theta_i|, 360 - |\theta_{DL} - \theta_k|) \leq k2$

           $[x_s, y_s, \vartheta_s, q_s] := (c_{DL}*[x_{DL}, y_{DL}, \vartheta_{DL}, q_{DL}] + c_k*[x_k, y_k, ,\vartheta_k, q_k])/(c_{DL} + c_k)$

           $m_{DL} := [x_s, y_s, \vartheta_s, q_s, c_{DL} + c_k]$

           $m_k := \{\}$

         end if

     end for

end for

Output: DL

6.5 Development of Unified Distance for 3D Minutiae Matching

The minutiae representation in 3D space requires 5 tuple values and each of these can be analysed to compute some unified or nonlinear distance for the matching. There are some interesting references [11] that have detailed the use of a function to compute the similarity score of 2D minutiae and usage of a threshold for the rejection of falsely matched minutiae. Assuming that there are some falsely matched 3D minutiae pairs, after the comparison with some hard threshold, it is possible to reject some of them by transforming all features to a scalar product. Therefore, it is judicious to develop a unified matching distance function for matching two 3D minutiae which can reject falsely matched minutiae using the comparison with a fixed decision threshold. This decision threshold can be computed offline during the system calibration or during the training stage.

The motivation explained earlier requires us to study the variation of five-tuple 3D minutiae components with the distance from the origin or the reference 3D minutia which generates a best match score for the genuine matches. Figure 6.12 presents the graphical illustration of the value of A_g, A_s, A_θ, cos N, r and A_ϕ, of the matched minutiae pair matching scores with different percentile against the distance from the reference minutiae. The values shown in these figures are sampled using a sliding window (distance -25, distance $+25$) which illustrates the relationship between (A_g, A_s, A_θ, cos N, r, A_ϕ) and distance from the 3D reference minutia. The trends shown in Fig. 6.11 suggest that the percentile values of cos$N$, ΔA_g, ΔA_s and ΔA_θ are relatively stable for smaller distances from the reference 3D minutia. The percentile values of Δr (labelled as *dist* on the respective figure caption), and ΔA_ϕ (labelled as A_{phi} on the respective figure caption) suggest some dependence or the relationship with the distance, especially near the region of interest (50–110), from the reference 3D minutia. Therefore, the thresholds for Δr and ΔA_ϕ f can be set as a function of the distance.

This study to combine the difference vectors for 3D minutiae, for generating a unified matching distance, suggests that Eqs. (6.11)–(6.15) can be generalized to more accurately account for relative variations in 3D feature distances as follows:

$$\text{funRSG}(\Delta v) = \left(\frac{\Delta r}{f(r)}\right)^a + \left(\frac{\Delta A_s}{A}\right)^b + \left(\frac{\Delta A_g}{B}\right)^c + \left(\frac{\Delta A_\theta}{C}\right)^d + \left(\frac{\Delta A_\phi}{f(r)}\right)^e + \left(\frac{1 - \cos N}{D}\right)^f$$

(6.17)

where Δv is the vector of difference values as computed in (6.10) and cos N is the value of normal vector (Fig. 6.3) cosine similarity between two matched minutiae. Above equation has an independent set of power term $\{a, b, \ldots, f\}$ while $f(r)$ can be some function of distance. The matching score between two 3D minutiae template P and Q can be computed from Eq. (6.16) using the number of matched minutiae pairs.

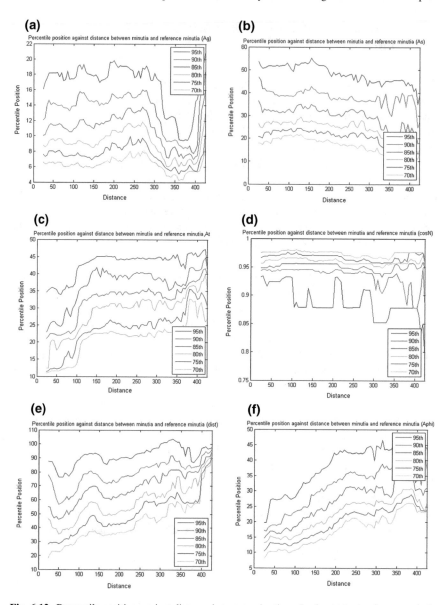

Fig. 6.12 Percentile position against distance between minutia and reference minutia respectively for **a** A_g, **b** A_s, **c** A_θ, **d** $\cos N$, **e** r, and **f** A_ϕ

6.6 Performance Evaluation

Lack of publicly available database from 3D fingerprint image is one of the key limitations to advance further research in this area. Therefore, new databases for 3D fingerprint images, using photometric stereo methods discussed in Chaps. 3–4, were developed [12, 16]. The contactless 3D fingerprint database in [15, 16] has been acquired from 240 distinct clients; by 'client', here, we refer a distinct finger, even if it belongs to the same person. We first acquired six images (impressions) from each of the fingers/clients which resulted in a total of 1440 impressions or 1440 3D fingerprint images reconstructed from 10,080 2D fingerprint images. The imaging hardware was developed to generate illuminations from seven different LEDs while acquiring the 2D image from the corresponding illumination for each of the fingerprint impressions. This entire 3D fingerprint database is now publicly accessible, along with image calibration (pixel positions) data along with source images, [16] to further research efforts in 3D fingerprint reconstruction and the matching.

Each of the acquired 2D fingerprint images is downsampled by four and then used to automatically extract 500×350 pixels size region of interest. This is achieved by subjecting the acquired images to an edge detector and then scanning the resulting image from boundaries to locate image centre which is used to crop fixed size region of interest images as discussed earlier. There are several 2D minutiae matching algorithms available in the literature [3, 7, 13] and our performance evaluation used BOZORTH3 [7] public implementation from NIST. Six 3D fingerprint images reconstructed from each of the 240 clients generated 3600 (240×15) genuine and 2,064,960 ($240 \times 6 \times 239 \times 6$) impostor matching scores. The average number of 3D minutiae recovered from the 3D fingerprint images was 34.64 while the average number of 2D minutiae per 2D fingerprint image, acquired for the 3D fingerprint reconstruction, was 40.28.

The method of 3D reconstruction can also influence the matching performance or the accuracy of recovered features. In case of photometric stereo based acquisition, the Poisson solver generates direct analytical results to the least square problem by solving a Poisson equation and has been shown to generate 3D fingerprint surface which has close resemblances to its natural shape. Our experiments for matching recovered 3D minutiae from the 3D fingerprints reconstructed using Poisson solver achieved superior performance than those from the 3D fingerprints reconstructed using Frankot–Chellappa algorithm discussed in Chap. 3. Figure 6.13 illustrates such comparative results for matching the recovered 3D minutiae from first 10 clients' 3D fingerprints. Therefore, this solution was preferred for matching 3D fingerprints using 3D minutiae in further experiments.

The experiments performed for the performance evaluation generated matching results from 240 client's 3D fingerprint templates. As argued earlier, multiple 2D fingerprint images acquired for the 3D fingerprint reconstruction can themselves be utilized for generating matching scores from the 2D fingerprint minutiae. However, the nature of imaging employed requires that each of these be acquired under different illumination, therefore we refer them as noisy 2D fingerprint images, as the 3D *shape from shading* is the key to reconstruct 3D fingerprint information. Therefore, the minutiae features extracted from the respective *noisy* 2D fingerprint images can be different and we attempted to generate best matching scores by matching all the available 2D fingerprint images from two client fingers. Figure 6.14a illustrates the experimental results from the usage of 2D fingerprint images when all such seven images corresponding to the query 3D fingerprint are matched with all the possible

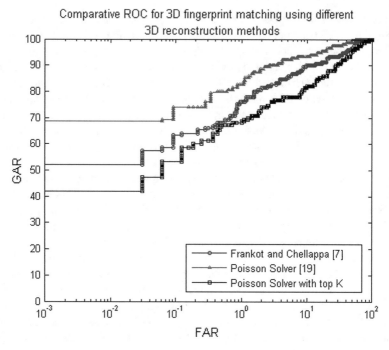

Fig. 6.13 Comparative matching accuracy from the 3D fingerprints reconstructed using different least square solutions

matches of the seven fingerprint are matched with all the possible matches of the images from the corresponding *probe* 3D fingerprint and using the best matching score as the decision score (for genuine or impostor classification) from the 2D fingerprint matching. As shown from the experimental results using receiver operating characteristics (ROC) in Fig. 6.14a, such an approach generates superior performance as compared to the case when only the best performing 2D fingerprint matching score is employed as the decision score. Therefore in our further experiments, we employed this superior approach to generating 2D fingerprint matching scores, corresponding to the reconstructed 3D fingerprint, by using all respective 2D fingerprint images utilized for the 3D fingerprint reconstruction. Figure 6.14b illustrates the distribution of (normalized) genuine and impostor matching scores obtained from 3D fingerprint and the corresponding 2D fingerprint matching.

In Sect. 6.5, a new function to combine the difference vector in (6.11)–(6.15) for generating unified matching distance was introduced. We also performed experiments to ascertain the performance and the employed nonlinear function (6.18) can be written as follows:

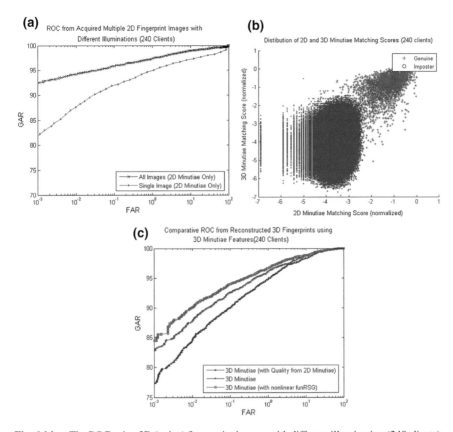

Fig. 6.14 **a** The ROC using 2D (noisy) fingerprint images with different illumination (240 clients), **b** distribution of matching scores, **c** comparative performance from 3D minutiae matching strategies considered in experiments

$$\text{funRSG}(\Delta v) = \left(\frac{\Delta r}{65}\right)^{0.8} + \left(\frac{\Delta A_s}{30}\right)^{0.8} + \left(\frac{\Delta A_g}{15}\right)^{0.8} + \left(\frac{\Delta A_\theta}{18}\right)^{0.8}$$
$$+ \left(\frac{\Delta A_\phi}{42}\right)^{0.8} + \left(\frac{1 - \cos N}{0.075}\right)^{0.8} \tag{6.18}$$

If funRSG(Δv) is smaller than 1.825, then this pair of 3D minutia is considered as the matched pair. The power term 0.8 makes the boundary of high dimension space convex. Figure 6.14c illustrates the comparative ROC generated from 3D fingerprint matching scores using two matching schemes. These results suggest that the unified score-based matching approach (6.18) achieves superior performance and is therefore employed in our further experiments. We also employed the 2D minutiae quality [7], corresponding to the matched 3D minutiae during score generation in (6.16) and attempted to achieve performance improvement. However, as can also be seen from the results in Fig. 6.14b, such an approach was not successful. This can be possibly attributed to the fact that 2D minutiae quality may not be a reliable indicator for 3D minutiae quality and an independent indicator needs to be developed for 3D minutiae quality in further work.

The ROC from matching 240 clients' images is shown in Fig. 6.15. This figure illustrates matching results using 3D minutiae representation and 3D fingerprint curvature representation. We also implemented 3D fingerprint matching approach described in [2] using the depth information. The ROC using this approach is also shown in Fig. 6.14a for comparison. As can be observed from this ROC, the resulting performance (EER of 18.56%) is quite poor. Our database of 3D fingerprints (or 3D model of reconstructing the 3D information) illustrates *larger distortion* after flatting the images than those in [2] from structured lighting approach. Significant degradation in performance (Fig. 6.15a) from the depth information matching can be attributed to such distortion in the flatting of the 3D fingerprint model. Our 3D fingerprint model does not reconstruct fingerprints as cylinder-like (because our reconstructed area/volume sometimes is partial or not cylinder-like), therefore the observed distortion is larger since the model in [2] assumes that the 3D fingerprints can be segmented into slices like from the cylindrical portions. Our efforts in Chap. 3 also investigated to reconstruct 3D fingerprints using non-Lambertian reconstruction model and the results were poor or not encouraging. The experimental results in Fig. 6.15a suggest that the 3D minutiae representation and matching approach developed in this work (Sects. 6.4 and 6.5) achieves superior results as compared to those using 3D fingerprint surface curvature representation.

Figure 6.15b illustrates experimental results from the combination of fingerprint matching scores using 2D minutiae representation (using the images acquired/available during 3D fingerprint reconstruction), 3D surface curvature representation, and 3D minutiae representation. This figure also illustrates matching results from corresponding 2D fingerprint images (reproduced from Fig. 6.13a) employed during the 3D fingerprint reconstruction. Table 5.2 presents comparative summary

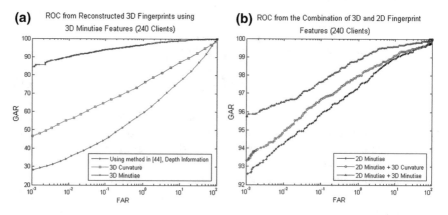

Fig. 6.15 The receiver operating characteristics for the **a** relative/comparative performance using reconstructed 3D fingerprint images, and **b** performance using combination of 3D fingerprint and 2D fingerprint images acquired during photometric stereo based reconstruction

Table 6.2 Individual and combined match performance from 2D and 3D fingerprint images

Experiments	2D minutiae (%)	3D curvature (%)	3D minutia (%)	2D minutiae + 3D curvature (%)	2D minutiae + 3D Minutiae (%)
Equal error rate from 240 clients (ROC in Fig. 6.15)	2.12	12.19	2.73	1.73	*1.02*
Equal error rate from 240 clients and 20 unknowns (DET in Fig. 6.16)	5.44	32.26	9.28	4.36	**3.33**
Rank-one accuracy from 240 clients and 20 unknowns (CMC in Fig. 6.17)	94.56	68.21	90.72	95.64	**96.67**

of equal error rate and rank-one recognition accuracy from the experimental results shown in Figs. 6.15–6.17. The score-level combination of matching scores using *adaptive fusion* [14] is employed in these experiments as it is judicious to exploit 3D matching scores only when the matching scores from 2D fingerprints are below some predetermined threshold. The threshold limit (Fig. 6.15b) for 3D minutiae was empirically fixed to 0.1 while this limit while combining 3D surface curvature was fixed to 0.09. It can be ascertained from this figure that the combination of 3D minutiae matching scores with the available 2D minutiae matching scores achieves superior performance. This performance improvement is significant and suggests that the 2D fingerprint images utilized for reconstructing 3D fingerprints, using photometric stereo, can be simultaneously used to improve the performance for the 3D fingerprint matching.

Any automated biometric system is also expected to effectively identify the unknown clients, i.e. able to reject those clients which are not enrolled in the database. In order to explicitly ascertain such capability, we additionally acquired 20 new clients' 3D fingerprint images and employed them as unknown clients. These unknown clients were then identified from the proposed approach to ascertain the performance. Figure 6.16b shows the plot of number of unknown clients identified as unknown versus known clients rejected as unknown. These experimental results also suggest superior performance for the 3D minutiae representation and achieve further improvement with the combination of conventional 2D minutiae features. The performance from the proposed identification schemes for the FPIR (false positive identification rate) and FNIR (false negative identification rate) was also observed and is illustrated in Fig. 6.16a. The performance improvement using the combination of 3D minutiae representation, extracted from reconstructed 3D fingerprint images, and 2D minutiae representation is observed to be quite consistent in FPIR versus FNIR plots.

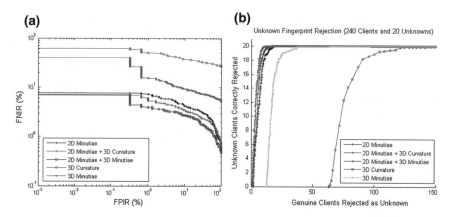

Fig. 6.16 **a** FPIR versus FNIR characteristics from the experiments and **b** corresponding perfor-mance for the unknown subject rejection using 240 clients and 20 unknowns

Although the key objective the performance evaluation was to ascertain perfor-mance from 3D fingerprint verification approach, we also performed the experiments for the recognition tasks using the same protocol/parameters as used for Fig. 6.13 for the verification task. Figure 6.17 illustrates cumulative match characteristics (CMC) from the recognition experiments for the comparison and comparison and combina-tion of 2D/3D fingerprint features. It is widely believed [17] that the verification and recognition are two different problems. However, the illustrated results for recogni-tion also suggest superior performance using 3D minutiae representation, over 3D curvature representation, and also illustrate the improvement in average (rank-one) recognition accuracy using combination of available minutiae features from the 2D fingerprint images acquired during the 3D fingerprint reconstruction. The score-level combination shown in Fig. 6.15b was also attempted with other popular methods and these results are shown in Fig. 6.18. The performance from other popular fusion approaches is also quite close and can also be employed to improve performance for 3D fingerprint identification using simultaneously available/acquired 2D fingerprint images.

6.7 Summary and Conclusions

This chapter introduced five-tuple representation of minutiae features in 3D spaces. Recovery of 3D minutiae, from the generalized 3D fingerprint data, to generate 3D minutiae fingerprint template was also discussed in section while systematic match-ing of such 3D fingerprint template was discussed in Sect. 6.4. The experimental results in the previous section support the usefulness of the developed methodology and importance of simultaneously available/acquired contactless 2D fingerprints in further improving the matching accuracy from 3D fingerprint images. The experi-

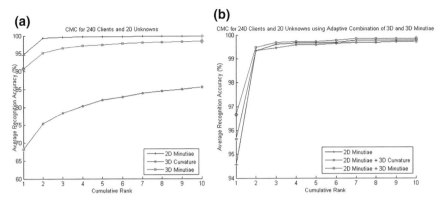

Fig. 6.17 The cumulative match characteristics for the **a** average recognition performance on using reconstructed 3D fingerprint images and **b** respective performance using combination of 3D fingerprint and 2D fingerprint images

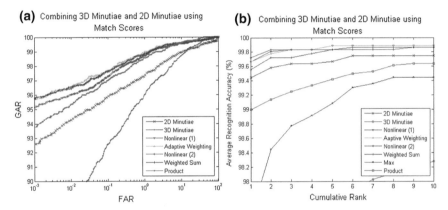

Fig. 6.18 The ROC (**a**) and CMC (**b**) from the combination of 3D fingerprint matching scores and 2D fingerprint matching scores using different fusion rules

mental results and the analysis in Sect. 6.4.2 underlines the need to develop quality measure for 3D minutiae. Such development of 3D fingerprint image quality measures would first require defining or standardizing the image resolution for intended applications, just like existing 500 dpi (level 2) or 1000 dpi (level 3 features) for conventional 2D fingerprint images. One simplified approach, which will also support interoperability between 3D and 2D fingerprints, would be to automatically compute these resolutions from the corresponding 2D images, e.g. 3D resolution of source or cloud point data whose projection on x–y plane would meet standardized 500 or 1000 dpi resolution.

References

1. Parziale G, Diaz-Santana E, Hauke R (2006) The surround imager: a multi-camera touchless device to acquire 3d rolled-equivalent fingerprints. In: Proceedings of ICB 2006, LNCS, vol 3832
2. Wang Y, Hassebrook LG, Lau DL (2010) Data acquisition and processing of 3-D fingerprints. IEEE Trans Info Forensics Secur 750–760
3. Maltoni D, Maio D, Jain AK, Prabhakar S (2009) Handbook of fingerprint recognition, 2nd edn. Springer, Berlin
4. Kumar A, Kwong C (2012) A method and device for contactless 3D fingerprint identification. Provisional U.S. Patent Application No. 61/680,716, 8 Aug 2012
5. Kumar A, Kwong C (2015) Contactless 3D biometric feature identification system and method thereof. U. S. Patent No. 8953854
6. Chen Y (2009) Extended feature set and touchless imaging for fingerprint matching, Ph.D. thesis, Michigan State University
7. NIST Biometric Image Software, NBIS Release 4.1.0, http://www.nist.gov/itl/iad/ig/nbis.cfm (2011)
8. Lalonde JF, Vandapel N, Hebert M (2006) Automatic three-dimensional point cloud processing for forest inventory. Technical Report No. MU-RI-TR-06-21, Robotics Institute, Pittsburgh
9. http://en.wikipedia.org/wiki/Atan2. Accessed Jun 2018
10. https://github.com/lessandro/nbis/blob/master/mindtct/src/lib/mindtct/quality.c (2018)
11. Liu L, Yang J, Zhu C, Jiang T (2006) Information theory based fingerprint matching. IEEE Trans Image Process 1100–1110
12. The Hong Kong Polytechnic University 3D Fingerprint Images Database Version 2.0, http://www.comp.polyu.edu.hk/~csajaykr/3Dfingerv2.htm (2017)
13. Tico M, Kuosmanen P (2003) Fingerprint matching using an orientation-based minutiae descriptor. IEEE Trans Pattern Anal Mach Intell 28(8):1009–1014
14. Kanhangad V, Kumar A, Zhang D (2011) A unified framework for contactless hand verification. IEEE Trans Info Forensics Secur 1014–1027
15. Kumar A, Kwong C (2015) Towards contactless, low-cost and accurate 3D fingerprint identification. IEEE Trans Pattern Anal Mach Intell 37:681–696
16. The Hong Kong Polytechnic University 3D Fingerprint Images Database, http://www.comp.polyu.edu.hk/~csajaykr/myhome/database.htm (2015)
17. Bolle RM, Connell JH, Pankanti S, Ratha NK, Seniorr AW (2005) The relation between ROC and the CMC, Proc. AutoID 2005, pp. 15–20

Chapter 7
Other Methods for 3D Fingerprint Matching

Matching of contactless 3D fingerprints can be performed by several methods, other than 3D minutiae templates. These methods of 3D fingerprints are specific to the nature of features extracted from the 3D fingerprints. In this chapter, we discuss three methods that are based on (a) surface curvature, (b) surface normal and (c) minutiae matching. These methods can be attractive for several online applications as they are computationally simpler.

7.1 Fast 3D Fingerprint Matching Using Finger Surface Code

Many discriminative 3D fingerprint features can be efficiently extracted and matched from 3D ridge–valley depth details. The formation of contactless 2D fingerprints images is highly dependent on the surface reflection properties, surface orientation, illumination and the sensor noise. In this context, the 3D fingerprint depth images are expected to be more reliable since the physical depth details, acquired using range sensing or structured lighting, are not expected to be influenced by the pose and illumination variations. Several references have illustrated that a range of binarized features, like barcode which is similar to iris code, which can be efficiently extracted/matched from the 3D depth images can offer very promising alternative for matching 3D fingerprints and is discussed in this section.

The shape index [1] discussed in Chap. 5 can be binarized and used to extract 3D fingerprint ridge–valley details as discussed in the previous chapter. It can describe local shape information and is particularly useful to describe nine well-known surface types [2]. Such binarized template representation, referred to as the surface code, was introduced in [3] to efficiently match 3D palmprint images. This *surface code* representation of 3D point cloud data attempts to adopt the surface-type information and encodes every surface pixel into one of the nine surface types.

© Springer Nature Switzerland AG 2018

A. Kumar, *Contactless 3D Fingerprint Identification*, Advances in Computer Vision and Pattern Recognition, https://doi.org/10.1007/978-3-319-67681-4_7

When the shape index C_i (Eq. 5.8 in Chap. 5) is close to 0.75, the shape of the 3D surface is more likely to be the ridge shape. On 3D fingerprint surface, the C_i's were observed to be concentrated in numeric values representing fingerprint valley (0.25) and ridge (0.75) regions. The surface index is therefore likely to be largely distributed in this zone. Therefore, our encoding scheme splits the 3D fingerprint surface into *five* zones: cup, rut, saddle, ridge and cap. The direction of the dominant principle curvature (max($|k_{max}|$, $|k_{min}|$) is portioned into six directions. Rut and ridge zones are further divided since cup, saddle and cap's $|k_{max}|$ and $|k_{min}|$ are close; therefore, t_{max} and t_{min} (defining the shape index C_i) are not as accurate as those in rut and ridge zones. The resulting feature representation has 15 different values. Each of these 15 values can therefore be represented using a 4-bit binary representation and thus forms a *binary code* for each pixel. Table 7.1 provides a summary of 3D surface curvature encoding using different shape index values and the corresponding 15 values that are encoded in 4-bit binary numbers for every 3D fingerprint surface points. This binarized representation of 3D fingerprint surface is referred to as *Finger Surface Code* [4] in this chapter and is similar to *IrisCode* representation in [5] or DoN in [6].

The matching score between two $M \times N$ *Finger Surface Codes*, say J and K, is computed using their normalized Hamming distance $HD(a, b)$ as follows:

$$\text{Score}_{\text{3D Finger}} = \frac{1}{4 \times M \times N} \sum_{p=1}^{N} \sum_{q=1}^{M} HD(J(p, q), K(p, q)) \qquad (7.1)$$

$$HD(a, b) = \begin{cases} 1, \text{ if } a \neq b \\ 0, \text{ if } a = b \end{cases} \qquad (7.2)$$

where $a, b \in \{0,1\}$. Two finger surface code templates are shifted left and right, and match score using (7.1) is generated for each of these shifts. The minimum score among all such shifts is designated as the final match score between two 3D Fingerprints.

Contactless 3D fingerprints from 136 different clients are used for the experimental evaluation. The finger surface codes were shifted by 51 bits to generate the best match score among two 3D fingerprints. Figure 7.1 shows the distribution of match scores using 3D finger surface codes and also using surface codes introduced in [3]. The distributions shown in this figure suggest that the finger surface code can further separate the genuine match scores and the impostor match scores. The comparative performance using the ROC is illustrated in Fig. 7.2. It can be observed from this figure that finger surface code-based approach for 3D fingerprint identification can offer significantly improved performance over those from surface codes.

The matching of 3D fingerprints using binary coding scheme, finger surface code, can be further enhanced by incorporating improved matching strategy. Such improved matching scheme [7] revisits the Hamming distance in (7.2) which is quite effective when all the values in the binarized code encode equally important information for discriminative identities. However, whether all the values in the coding space

Table 7.1 The zones of the *Finger Surface Code*

C_i	0–0.0625	0.0625–0.4375					0.4375–0.5625	0.5625–0.9375					0.9375–1
Angle(pi/6)	–	0	1	2	3	4	5	/	0	1	2	3	4	5	–
Code	0	0	1	2	3	4	5	6	7	8	9	10	11	12	13	14

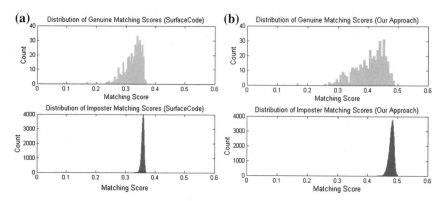

Fig. 7.1 Distribution of genuine and impostor match scores for 136 clients 3D fingerprints using (**a**) surface code and (**b**) finger surface code

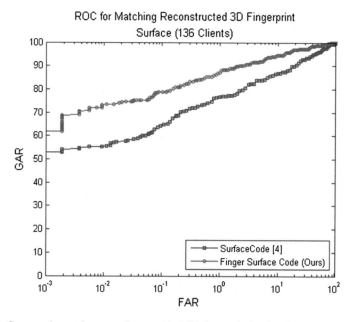

Fig. 7.2 Comparative performance for matching 3D fingerprints using finger surface code and surface code for 136 clients 3D fingerprints

encode equally important information depends on the choice of feature extraction and discretization/binarization methods. Therefore, such equal probability or maximum entropy assumption may not be always true. For example, when the Hamming distance is used for *Surface Code*, distance between level 1 (0001) and level 5 (0101) is 1, while the distance between level 7 (0111) and level 8 (1000) is 4. These resulting distances cannot correctly represent the actual difference between different levels. This limitation also exists in the *Finger Surface Code*. Therefore, [7] details a more effective similarity measure SM(a, b), to replace Hamming distance (7.2) HD(a, b), that can judiciously consider the individual importance of features in the coding space for computing the matching scores.

$$SM(a, b) = \begin{cases} 2 - s, & \text{if } a = b = 1 \\ s, & \text{if } a = b = 0 \\ 0, & \text{if } a \neq b \end{cases} \tag{7.3}$$

The parameter s in above equation can control the significance of one of the coding pairs. Hamming distance is a special case when s is set to be 1. If the four possible scenarios ($ab \in \{00,01,10,11\}$) are equally likely, the expected similarity score will be 0.5, which is independent of the parameter s. The experimental results using this approach ($s = 0.75$), referred to as *efficient finger surface code*, are shown in Fig. 7.3. These experiments use 3D fingerprints from 240 different clients [8] for matching comparative performance evaluation. The 3D fingerprints from 240 clients, each with six images, resulted in 3600 (240 × 15) genuine and 1,032,480 (239 × 6 × 6 × 240/2) imposter match scores. In order to account for the translation variations in this database, the templates are shifted with vertical and horizontal translations. The minimum score obtained from such shifting is considered as the final match score. Comparative performance shown in Fig. 7.3 indicates that such matching strategy can further improve the matching accuracy and offers efficient alternative for matching 3D fingerprint images.

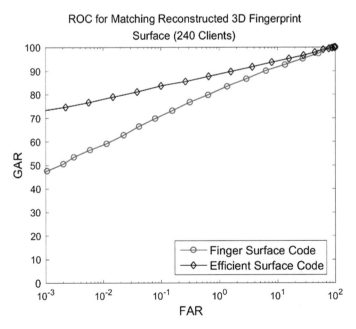

Fig. 7.3 Comparative experimental results for 3D fingerprint matching from 240 different clients with different 3D fingerprints

7.2   Tetrahedron-Based 3D Fingerprint Matching

We discussed the generation of 3D fingerprint minutiae templates in the previous chapter. Matching these 3D minutiae templates required complex transformations, as discussed in the last chapter, for every minutia to align them in 3D space. This was the main reason for higher computational complexity in matching such templates. This complexity can be significantly reduced for matching 3D minutiae templates using Delaunay tetrahedron-based alignment and matching. Delaunay triangulation approach has been evaluated for conventional 2D fingerprint identification in the literature [9, 10]. Such triangulation is formed from localized 2D fingerprint minutiae, in the respective templates, using a unique topological structure. As compared to number of possible minutiae triangles under a given number of minutiae, the Delaunay triangulation significantly reduces the number of minutiae triangles. This reduction significantly helps in reducing computational complexity for topological matching using graph matching-based approaches. The Delaunay triangulation can be efficiently computed by first generating Voronoi diagram as the Delaunay triangle corresponds to dual graph of Voronoi diagram [11]. Delaunay triangulation from the localized minutia features is used to recover scale-invariant features as shown in Fig. 7.4a. Each of such triangles can be used to compute the following three features:

$$0 \le \frac{l_1}{l_3} \le 1, \ 0 \le \frac{l_2}{l_3} \le 1, \ -1 \le \cos(\varphi) \le 1 \tag{7.4}$$

where l_1 represents the ith side of the minutiae triangle and the lengths of three sides are sorted in ascending order, i.e. $l_1 \le l_2 \le l_3$ φ represents the largest angle of such minutia triangle and m_1 represents the ith minutia. Features are first extracted from these triangulations and then minutiae alignment and fingerprint matching are performed. A 3D space, the Delaunay triangulation generates tetrahedra that satisfies the empty circumsphere criterion, similar to empty circumcircle criterion in 2D case. The key idea is to generate Delaunay tetrahedrons based on the minutiae m_i in 3D space. For simplicity, the term tetrahedron is used here to represent 3D Delaunay tetrahedron. A general tetrahedron is defined as a convex polyhedron consisting

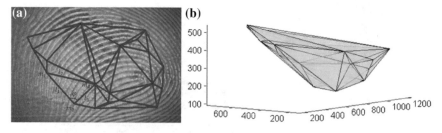

Fig. 7.4 **a** 2D minutiae Delaunay triangulation and features. li represents 2D minutia, presents the side of minutiae triangle, presents the largest angle of minutiae triangle. **b** The triangulation of a 3D minutiae is composed of tetrahedra

of four triangular faces. It fills the convex hull of the points with tetrahedron so that the vertices of the tetrahedron are those of the data points, i.e. minutiae. It can be specified by its polyhedron vertices as (x_i, y_i, z_i) where $i = 1, \ldots, 4$. The circumscribing sphere of any tetrahedron does not contain any other point inside sphere. The algorithm of generating such tetrahedron is detailed in [12]. Every four 3D minutiae are connected and generate one tetrahedron. Each of the four minutiae corresponds to tetrahedron's four vertices (m_1, m_2, m_3, m_4). Each vertex can be represented as $m = [x, y, z, \theta, \phi]$. The tetrahedron can be uniquely represented from the following 8-tuple representation in 3D space:

$$\left[l_{max}, l_{min}, \frac{l_{max}}{l_{s\_max}}, \varphi, \tilde{\theta}_{max} \tilde{\theta}_{min}, \tilde{\phi}_{max}, \tilde{\phi}_{min} \right] \qquad (7.5)$$

where l_{max} and l_{min} present the largest side and smallest side of the tetrahedron, respectively. l_{s_max} is the second largest side, while φ is the largest angle of each tetrahedron's four faces. The length of any tetrahedron side can be computed as follows:

$$l = \sqrt{(x_1 - x_2)^2 + (y_1 - y_2)^2 + (z_1 - z_2)^2} \qquad (7.6)$$

Besides these geometric features, the differences of minutiae direction and orientation can be computed as features.

$$\tilde{\theta} = \theta_1 - \theta_2 \qquad (7.7)$$
$$\tilde{\phi} = \phi_1 - \phi_2 \qquad (7.8)$$

$\tilde{\theta}_{max}$ computes the 2D orientation difference between two vertices of the largest side in tetrahedron, i.e. azimuth angle difference between two 3D minutiae connecting largest side l_{max}, $\tilde{\phi}_{max}$ is the 3D orientation difference between two vertices of the largest side in tetrahedron, i.e. elevation angle difference between two 3D minutiae connecting the largest side l_{max}. These eight features describing a tetrahedron in (7.5) are employed to match two arbitrary Delaunay tetrahedrons. For each matched tetrahedron, we perform 3D alignment based on its vertex, i.e. 3D minutiae. Figure 7.5a illustrates a 3D minutiae tetrahedron, and 7.5b illustrates one of its minutia representation in 3D space. This sample figure in (a) uses m_1, m_2, m_3, m_4 to represent the four 3D minutiae, l_{max} and φ represents the two tetrahedron features (7.8), while the sample figure in (b) defines measurements for one of the 3D minutia m_3 of this tetrahedron with θ_3 and ϕ_3 being the 2D and 3D orientation of this 3D minutia m_3. Redline in this figure is the projection of 3D minutia orientation ϕ_3 (blue line) on the x–y plane.

The 3D minutiae tetrahedron matching algorithm can generate a numerical match score between the two 3D fingerprint minutiae templates. A reference minutia tetrahedron sample TP_i and probe minutia tetrahedron sample TQ_j are, respectively, selected from the 3D fingerprint templates P and Q. The features in (7.8) are computed from

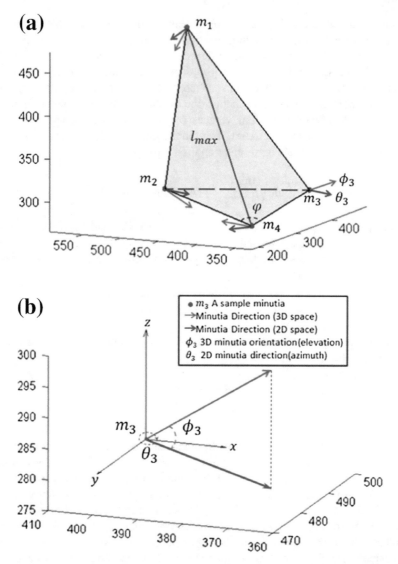

Fig. 7.5 a A tetrahedron sample and its features: m_i represents 3D minutia, l_{max} in red line represents the largest side in this tetrahedron, φ represents the largest angle of the tetrahedron's face. **b**
Minutiae sample of tetrahedron and its minutiae direction θ_3 and orientation ϕ_3 (blue line). The red
colour line is the projection of blue colour line on x–y plane, representing orientation of m_3 minutia
and is used to illustrate the measurement of angle θ_3 representing minutiae direction

these two minutiae tetrahedron samples. If the difference between the TP_i and TQ_j
features is smaller than a given threshold, these two minutiae tetrahedron samples
can be considered as being matched.

$$\Delta l\text{max} = \left|l_{P_i\text{max}} - l_{Q_j\text{max}}\right| \Delta\text{max} = \left|l_{P_i\text{max}} - l_{Q_j\text{max}}\right|$$

$$\Delta l = \left|\frac{l_{P_i\text{max}}}{l_{P_{is}\_\text{max}}} - \frac{l_{Q_j\text{max}}}{l_{Q_{js}\_\text{max}}}\right|, \quad \Delta\varphi = \left|\varphi_{P_i} - \varphi_{Q_j}\right| \tag{7.9}$$

$$\Delta\tilde{\theta}\text{max} = \left|\tilde{\theta}_{P_i\text{max}} - \tilde{\theta}_{Q_j\text{max}}\right|$$

$$\Delta\tilde{\theta}\text{min} = \left|\tilde{\theta}_{P_i\text{min}} - \tilde{\theta}_{Q_j\text{min}}\right| \tag{7.10}$$

$$\Delta\tilde{\phi}\text{max} = \left|\tilde{\phi}_{P_i\text{max}} - \tilde{\phi}_{Q_j\text{max}}\right|$$

$$\Delta\tilde{\phi}\text{min} = \left|\tilde{\phi}_{P_i\text{min}} - \tilde{\phi}_{Q_j\text{min}}\right| \tag{7.11}$$

when $\Delta l\text{max} < \text{th}_{l\text{max}}$, $\Delta l\text{min} < \text{th}_{l\text{min}}$, $\Delta l < \text{th}_l$, $\Delta\varphi < \text{th}_\varphi$, $\Delta\tilde{\theta}\text{max} < \text{th}_{\tilde{\theta}}$. $\Delta\tilde{\theta}\text{min} < \text{th}_{\tilde{\theta}}$, $\Delta\tilde{\phi}\text{max} < \text{th}_{\tilde{\phi}}$, $\Delta\tilde{\phi}\text{min} < \text{th}_{\tilde{\phi}}$, two tetrahedra are considered as being matched. Then, 3D minutiae tetrahedron alignment is performed from the transformation matrix from (6.9)–(6.10) of Chap. 6. The match score between two 3D fingerprints can be computed as follows:

$$S_{3DT} = \frac{m^2}{N_P * N_Q} \tag{7.12}$$

where m is the total number of matched 3D minutiae pairs, N_P and N_Q is the number of 3D minutiae in 3D fingerprint template P and Q. This algorithm for matching two 3D fingerprints is also summarized in the algorithm TM.

Algorithm TM: Tetrahedron based 3D Fingerprint Matching

Input: Two 3D fingerprint minutiae template P, Q with corresponding 3D minutiae m_{Pi} and m_{Qj};

Output: Match score between each 3D fingerprint minutiae templates;

1: Generate tetrahedron model TP and TQ using 3D minutiae, each tetrahedron is defined as a convex polyhedron consisting of four triangular faces which are specified by its polyhedron or 3D minutia vertexes (x_i, y_i, z_i) where $i = 1, \ldots 4$;
2: **for** each tetrahedron TP_i in TP **do**
3:   extract features using (5)-(8);
4:   **for** each tetrahedron TQ_j in TQ **do**
5:       extract features using (5)-(8);
6:       **if** equation (9), (10), (11) < Threshold **then**
7:           **for** each vertex V_{TPi} in TP_i and V_{TQj} in TQ_j **do**
8:               3D minutiae alignment and matching
        compute match score using (12)
9:           **end for**
10:       **end if**
11:   **end for**
12: **end for**

7.2.1 3D Minutiae Hierarchical Tetrahedron Matching

The tetrahedron-based 3D fingerprint matching achieves promising results with smaller alignment and the matching time. It should be noted that the spurious minutiae may introduce false tetrahedron and missing minutiae may remove the correct tetrahedron. This is likely to degrade the matching performance. In order to address this problem and to improve the match accuracy, we can attempt to introduce 3D minutiae *hierarchical* tetrahedron matching. In this approach, the 3D minutiae are divided into different classes based on the minutiae quality. This minutiae quality can be estimated from the surface curvature images of 3D fingerprints as discussed in Sect. 6.4.1 of Chap. 6. The 3D minutiae are then connected to generate hierarchical tetrahedrons, and these hierarchical tetrahedrons are employed for the matching. Figure 7.6 illustrates samples of tetrahedrons that have been classified into one of the three classes. In different classes, tetrahedron is generated from the minutiae with different minutiae quality ($q > 0.7$, $q > 0.5$ and all the minutiae). The duplicate tetrahedrons in different classes are ignored. The 3D minutiae tetrahedron matching is then employed to match these hierarchical tetrahedrons. As a result, relatively correct minutiae are more likely to be matched in the first class (Fig. 7.6). The missing minutiae are more likely to be matched in class three. The matching process of hierarchical tetrahedron method is similar to 3D minutiae tetrahedron matching algorithm discussed in the previous section. The match score is computed from all the different classes of tetrahedrons.

The experimental validation presented in this section employed public 3D fingerprint database in [13]. The first session database was acquired from 300 different client fingers and contains 3600 coloured 2D fingerprint images. For each of the client fingers, six 2D image samples were automatically acquired using the setup introduced in Chap. 4. This resulted in 1800 3D fingerprint images which were

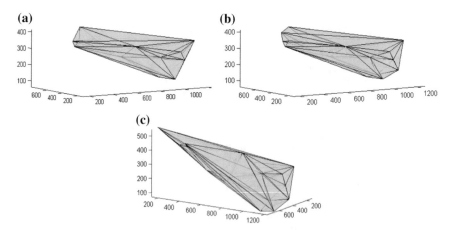

Fig. 7.6 **a** Tetrahedron with minutiae quality > 0.7, **b** Tetrahedron with minutiae quality > 0.5 **c** Tetrahedron with all minutiae

reconstructed using 10,800 (1800 × 6) greyscale 2D images that were generated from 1800 colour 2D fingerprint acquired under coloured photometric stereo. The second session images were acquired after an average interval of 33 weeks (maximum of 65 weeks and minimum of 7 weeks from a volunteer). However, not all volunteers were available for the second session 3D fingerprint imaging, and therefore second session database was acquired from 200 different client fingers with 2400 (200 x 2 x 6) coloured 2D fingerprint images. The image preprocessing, 3D reconstruction and fingerprint identification programs were run on a PC with i7-4770 CPU and 32 GB RAM. The original resolution of acquired 2D fingerprint images is 2048 × 1536. We used automatically segmented 1400 × 900 pixels region of interest images to reconstruct 3D fingerprint. The experimental results for verification experiments, using the first session 3D fingerprints, employed only the first 300 clients data and generated 4500 (300 × 15) genuine and 1,614,600 (300 × 6 × 6 × 299/2) imposter match scores. The experimental results using second session database generated 1200 (200 × 6) genuine and 238.800 (200 × 199 × 6) imposter match scores. The average number of 3D minutiae from 3D fingerprint in the first session data is 30.356 and 31.346 in second session data. Figure 7.7 illustrates comparative results using ROC from such first session and second session experiments. These results indicate that hierarchal tetrahedron-based approach can offer superior performance than matching using tetrahedron-based approach. The performance from second session 3D fingerprint comparisons is degraded as compared with those from only first session 3D fingerprint data. The key reason for such degradation is related to significant variations between the second session and first session 3D fingerprint images. Such high intra-class variation can be attributed to the positioning or the presentation of finger from the volunteers under contactless imaging setup. One possible approach to significantly reduce such intra-class variations is to employ finger-guide, as in [15], which can be considered as a tradeoff for performance as fixing locations of 3D finger using finger-guide can degrade advantages from the 'contactless' imaging. Average time complexity of matching two 3D fingerprints using 3D minutiae tetrahedron is 0.167 s and using 3D hierarchical tetrahedron is 0.361 s. This is significantly smaller than using 3D minutiae (2.435 s) approach discussed in the previous chapter. The comparative results shown in Fig. 7.7 should be considered preliminary, as further work is required to incorporate preprocessing steps for the alignment of two-session 3D fingerprints or to reduce intra-class variations before the matching. Our results indicate that the key benefit from the 3D minutiae tetrahedron-based 3D fingerprint matching lied in significant in reducing computational complexity while ensuring similar or marginally better performance over those from the direct usage of 3D minutiae templates.

7.3 3D Fingerprint Matching Using Surface Normals

Another approach to match 3D fingerprints considers [14] the unit normal vector location, at all the 3D fingerprint surface points or only sampled at 3D minutiae

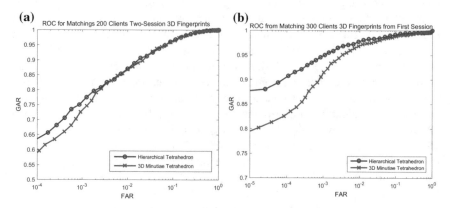

Fig. 7.7 Contactless 3D fingerprint matching results using the ROC; **a** using first session 3D fingerprint matching and **b** using two session fingerprint matching

locations. The unit normal vector at 3D fingerprint surface point can be estimated using lowest eigenvalues (Eq. 5.7 in Chap. 5) which are available while determining principal axes of the masked ridge surface for the minutiae direction angle ϕ. Since the direction of principal axis is the normal vector over the masked ridge surface, it has least noise as compared to the case when the normal vector is estimated/measured on the exact minutiae location. The match score between the two unit normal vectors, say n_1 and n_2 from the 3D minutiae of query image (input file) and that of the stored template file, is generated using their dot product, i.e. using $\cos(N) = n_1 \cdot n_2$. If $\cos(N)$ is bigger than a predefined threshold, then the normal vectors of the minutiae pair are considered to be matched. If $\cos(N)$ is larger than a predefined threshold, then the normal vector(s) from the two 3D fingerprints are considered to be matched. The match scores between the surface normal vectors from 3D fingerprints can provide additional cue and used to further improve the matching accuracy. Figure 7.8 illustrates such sample experimental results from 3D fingerprints of 10 different clients. The ROCs in this figure also presents the results using score-level combination from surface norms and finger surface codes. These results indicate that the 3D fingerprint matching accuracy can be further improved by incorporating surface normal match scores.

7.4   Summary and Conclusions

This chapter introduced computationally simpler alternatives for matching contactless 3D fingerprint images. The binarized surface codes generated from the 3D surface curvature features, and matched using the similarity measure in (7.3), offers most attractive alternative for fast 3D fingerprint identification in a range of civilian

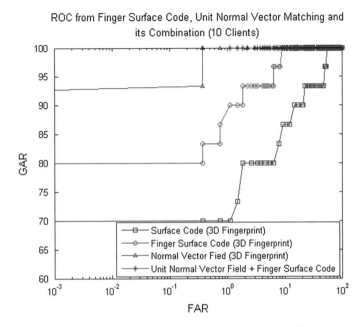

Fig. 7.8 Experimental results from unit surface normal vector field based 3D fingerprint matching and performance improvement using its adaptive combination finger surface code-based matching

applications. Tetrahedron based 3D fingerprint matching introduced in Sect. 7.2 is essentially an extension of Delaunay triangulation based 2D fingerprint matching in 3D space. This approach implicitly assumed that 3D minutiae templates are available and therefore adds to computations as compared with two other methods in Sects. 7.1 and 7.3. The 8-tuple representation in (7.5) has two terms, i.e., l_{max} and l_{min}, that are sensitive to scale changes and can degrade accuracy for the tetrahedron matches. The experimental results in Sect. 7.2 indicates that tetrahedron based 3D fingerprint matching can provide faster alternative to match two 3D fingerprint templates as it can avoid computationally complex 3D minutiae template alignment. As detailed in Sect. 7.3, Surface normals from the 3D fingerprint source data describes 3D shape and provide computationally simpler alternative to improve 3D fingerprint matching accuracy.

The user interaction with contactless 3D fingerprint imaging sensor can generate significant intra-class variations over those images acquired from conventional 2D fingerprint sensors. Such high intra-class 3D fingerprint variation is essentially cost or the tradeoff to avail high degree of freedom and benefits associated with contactless finger imaging. Such high intra-class variations are also reflected in two-session 3D fingerprint database in [13] and is the key reason for the significant degradation in performance (Fig. 7.7b) from matching two 3D fingerprints acquired from the same user/finger after a long interval. Matching 3D fingerprints with significantly large intra-class variations, i.e. in their elevation angles, azimuth angles and distances with sensor, can be considered as a partial 3D fingerprint matching problem [16] which is expected to degrade the matching accuracy. Acquisition of nail-to-nail 3D fingerprint images can significantly improve the matching accuracy for 3D fingerprints with large intra-class variations. Unlike for the partial-3D fingerprints generated from a fixed sensor, nail-to-nail 3D fingerprints can offer significantly higher image details to accurately match 3D fingerprints even under higher intra-class variations.

References

1. Koenderink JJ, van Doorn AJ (1992) Surface shape and curvature scales. Image Vis Comput (8):557–564
2. Dorai C, Jain AK (1997) COSMOS-A representation scheme for 3D free-form objects. IEEE Trans Pattern Anal Mach Intell 19(10):1115–1130
3. Kumar A, Kwong C (2013) Towards contactless, low-cost and accurate 3D fingerprint identification. In: Proceedings of the CVPR 2013, Portland, USA, June 2013, pp 3438–3443
4. Kanhangad V, Kumar A, Zhang D (2011) A unified framework for contactless hand identification. IEEE Trans Inf Forensics Secur 20(5):1415–1424
5. Daugman J (2003) The importance of being random: statistical principles of iris recognition. Pattern Recogn 36:279–291
6. Zheng Q, Kumar A, Pan G (2016) A 3D feature descriptor recovered from a single 2D palmprint image. IEEE Trans Pattern Anal Mach Intell 38:1272–1279
7. Cheng KHM, Kumar A (2018) Advancing surface feature description and matching for more accurate biometric recognition. In: Proceedings of the ICPR 2018, Beijing, Aug 2018
8. The Hong Kong Polytechnic University 3D Fingerprint Images Database (2015) http://www.comp.polyu.edu.hk/~csajaykr/myhome/database.htm
9. Uz T, Bebis G, Erol A et al (2007) Minutiae-based template synthesis and matching using hierarchical Delaunay triangulations. In: Proceedings of the BTAS, 2007, pp 1–8
10. Parziale G, Niel A (2004) A fingerprint matching using minutiae triangulation. In: Proceedings of the ICBA 2004, Hong Kong, Springer Berlin Heidelberg, pp 241–248
11. Ahuja N (1982) Dot pattern processing using voronoi neighborhoods. IEEE Trans Pattern Anal Mach Intell 4(3):336–343
12. Fang TP, Piegl LA (1995) Delaunay triangulation in three dimensions. IEEE Trans Comput Graphics Appl 15(5):62–69
13. The Hong Kong Polytechnic University 3D Fingerprint Images Database Version 2.0 (2017) http://www.comp.polyu.edu.hk/~csajaykr/3Dfingerv2.htm
14. Kumar A, Kwong C (2015) Contactless 3D biometric feature identification system and method thereof. U.S. Patent No. 8953854, Feb 2015
15. Galbally J, Bostrom G, Beslay L (2017) Full 3d touchless fingerprint recognition: sensor, database and baseline performance, Proc. IJCB 2017, pp. 225–232
16. Lin C, Kumar A (2018) Contactless and Partial 3D Fingerprint Recognition using Multi-view Deep Representation, Pattern Recognition, pp. 314–327, Nov. 2018

Chapter 8
Individuality of 3D Fingerprints

2D fingerprint impressions have been widely used to uniquely establish the identity of an individual for over hundred years. Several attempts have been made in the literature to quantitatively establish the *limits* on the number of different identities, i.e. uniqueness, that can be established from the 2D fingerprint images. Reference [1] provides critical analysis of many such fingerprint image individuality models introduced in the literature. These models can be classified into five different categories [2]: grid-based models, ridge-based models, fixed probability models, relative measurement models and generative models, and [3, 4, 5, 6 and 7] are examples of representative samples in each of these five categories, respectively. Studies on the individuality of fingerprint biometric have attracted renewed attention following several lawsuits in US courts that have challenged the admissibility of personal identification using fingerprint biometric-based evidences. These challenges have primarily questioned the uncertainty associated with the experts' judgment and quantitative estimation on the likelihood of erroneous decisions, which are generally made on the basis of false random matches between two arbitrary fingerprint samples from different individuals.

Uniqueness of fingerprint biometric has also been widely accepted by the experts on the basis of manual inspection from the 2D fingerprint images. However, any scientific study on the uniqueness of 3D fingerprints, or the merit of employing 3D fingerprint identification over the conventional 2D fingerprint identification, has not yet been performed. This chapter attempts to answer one of the most fundamental questions on the availability of inherent discriminable information from the 3D fingerprints. There are many choices for representing 3D fingerprint data. The fingerprint representation based on 3D minutiae introduced in the previous chapter is the most effective to quantitatively study the discriminability of 3D fingerprint biometric. The permanence of fiction ridges on finger surfaces, or the resulting minutiae features, has been supported by several morphogenesis and anatomical studies [8, 9]. Therefore, the key topic of interest in establishing the individuality of 3D fingerprints is to measure the amount of *discriminable* details in two different 3D fingerprint images

© Springer Nature Switzerland AG 2018
A. Kumar, *Contactless 3D Fingerprint Identification*, Advances in Computer Vision and Pattern Recognition, https://doi.org/10.1007/978-3-319-67681-4_8

using the minutiae features. The images in Fig. 8.1 indicate merit in employing 3D fingerprints to enhance uniqueness of fingerprint biometric.

There are many possibilities to ascertain the individuality of 3D fingerprints. In this work, we formulate the individuality problem [7] as the probability of finding sufficiently similar 3D fingerprint in a given population, i.e. given two 3D fingerprints with same resolution but originating from two *different* sources, what is the probability that these 3D fingerprints can be declared as sufficiently similar with respect to a given or popular matching criterion? In order to compute such probability of false random correspondence between two arbitrarily chosen 3D fingerprints, we will need to first select a matching criterion and the representation for 3D fingerprints. Contactless 3D fingerprints can be represented by many features and given the popular usage of minutiae features in the law enforcement, forensics and in most 2D fingerprint identification systems, it is judicious to choose 3D minutiae representation introduced in Chap. 6 to ascertain the individuality of 3D fingerprints. Any such scientific basis to ascertain the individuality of 3D fingerprints can help to determine the merit of employing 3D fingerprints, over conventional 2D fingerprints, for the

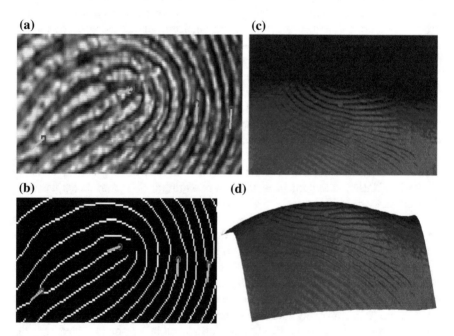

Fig. 8.1 **a** Processed 2D fingerprint image sample with two bifurcations (blue square) and two endings (red circle), along with the orientation (green line) marked on the image; **b** thinned images from 2D image in (**a**) with respective minutiae. **c** same fingerprint as in image (**a**) using 3D fingerprint imaging; **d** reconstructed 3D fingerprint from the same finger as in image (**a**) and with four of five minutiae marked in 3D image using blue colour spots. Acquisition of 3D fingerprints enables recovery of same 2D minutiae features with additional discriminative information, i.e. height and elevation angle, in 3D space that significantly enhances the limits on the individuality of fingerprint biometric

personal identification. Such measure of individuality can also be used to determine upper bound on the expected performance from the 3D fingerprint identification systems.

8.1 Probability of False Random Correspondence Between Two 3D Fingerprints

Let P and Q denote two arbitrarily selected 3D fingerprint minutiae templates using 3D minutiae representations, with m and n, respectively, representing the number of truly recovered 3D minutiae that are available for the matching.

$$P = \{(x_1, y_1, z_1, \theta_1, \phi_1), \ldots, (x_m, y_m, z_m, \theta_m, \phi_m)\} \tag{8.1}$$

$$Q = \{(x_1', y_1', z_1', \theta_1', \phi_1'), \ldots, (x_n', y_n', z_n', \theta_n', \phi_n')\} \tag{8.2}$$

After alignment as discussed in Chap. 6, the minutiae i in the template Q is considered as being matched to the minutiae j in template P if and only if it can meet the following predetermined criterion for the matching:

$$\sqrt{(x_i' - x_j)^2 + (y_i' - y_j)^2 + (z_i' - z_j)^2} \le r_0 \tag{8.3}$$

$$\min(|\theta_i' - \theta_j|, 360 - |\theta_i' - \theta_j|) \le \theta_0 \tag{8.4}$$

$$\min(|\phi_i' - \phi_j|) \le \phi_0 \tag{8.5}$$

where r_0, θ_0 and ϕ_0 represent the tolerance limit in measurements of the distance and tolerance limits in measurements of the angles, respectively. Let V be the overlapping volume between two arbitrary 3D fingerprint surfaces being matched, while $H(p)$ be the height of the aligned surface with point p on the projection on x–y plane. Figure 8.2 illustrates the spherical region of tolerance between two 3D minutiae matching regions within the overlapping volume of two matched 3D fingerprints. The V can be estimated as follows:

$$U = \sum_{s=1}^{N} \sum_{t=1, t \ne s}^{N} \frac{\sum_{p \in A} |H(p)_s - H(p)_t|}{A(s, t)} \tag{8.6}$$

$$\sigma = \sum_{s=1}^{N} \sum_{t=1, t \ne s}^{N} \frac{\left(\sum_{p \in A} (|H(p)_s - H(p)_t| - U)^2\right)}{A(s, t)} \tag{8.7}$$

$$V = (U + 2\sqrt{\sigma}) A(s, t) \tag{8.8}$$

where $A(s, t)$ is the overlapped area on x–y plane of two aligned surfaces s and t. The variance σ can be estimated by statistics of the reconstructed 3D fingerprint surface in the database and represents twice standard deviation limits that cover 95%

Fig. 8.2 The overlapping
volume between two imaged
D finger regions for two 3D
fingerprints. The blue and
red colour arrows illustrate
azimuth (θ) and elevation
(ϕ) angles for the minutiae
points. The tolerance
spherical region is illustrated
for a sample 3D minutia
using a dotted red colour
circle with radius r_0

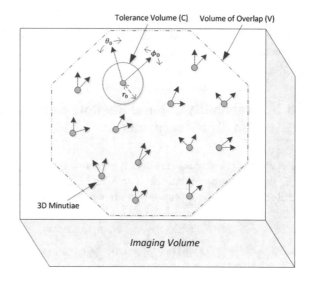

confidence interval. The probability of 3D minutiae in a template matched with an
arbitrary minutia is given by

$$P\left(\sqrt{(x_i' - x_j)^2 + (y_i' - y_j)^2 + (z_i' - z_j)^2} \leq r_0\right) = \frac{\text{volume of tolerance}}{\text{volume of overlap}} = \frac{\frac{4}{3}\pi\, r_0^3}{V} = \frac{C}{V} \quad (8.9)$$

$$P\left(\min(|\theta_i' - \theta_j|, 360 - |\theta_i' - \theta_j|) \leq \theta_0\right) = \frac{\text{angle of tolerance}}{\text{total angle}} = \frac{2\theta_0}{360} \quad (8.10)$$

$$P\left(\min(|\phi_i' - \phi_j|) \leq \phi_0\right) = \frac{\text{angle of tolerance}}{\text{total angle}} = \frac{2\phi_0}{180} \quad (8.11)$$

assuming (x, y, z) and (x', y', z') are independent in (8.9), θ and θ' are independent
in (8.10), and ϕ and ϕ' are independent. Let $J = \frac{V}{C}$, and assuming J is the nearest
integer.

Let us first estimate the probability that two arbitrary 3D minutiae can have correspondence, or be matched, in 3D space and later incorporate the additional requirements for these minutiae to be fully considered as matched for their 3D spatial
orientations. Given two arbitrary and two 3D fingerprint templates, the probability
that *only one* 3D minutia from the input template can be falsely matched with any
of the m minutiae in the target (or the second) 3D fingerprint template is

$$P_{3D}(1, m, 1) = \frac{mC}{V} \quad (8.12)$$

Extending (8.12), we can now estimate the probability that *only two* 3D minutiae
from the input template can be falsely matched with any of the m minutiae in the
target 3D fingerprint template:

$$P_{3D}(2, m, 1) = P(\text{first minutiae corresponded but second not})$$
$$+ P(\text{second minutiae corresponded but first not}) \tag{8.13}$$

$$P_{3D}(2, m, 1) = \left(\frac{mC}{V}\right)\left(\frac{V - mC}{V - C}\right) + \left(\frac{mC}{V}\right)\left(\frac{V - mC}{V - C}\right)$$

$$= 2\left(\frac{mC}{V}\right)\left(\frac{V - mC}{V - C}\right) \tag{8.14}$$

We can similarly extend (8.14) to estimate the probability that any of the n minutiae from input templates can be matched/corresponded with any of the m template minutiae in the target template as follows:

$$P_{3D}(mn, 1) = n\left(\frac{mC}{V}\right)\left(\frac{V - mC}{V - C}\right)$$

$$= \binom{n}{1}\left(\frac{mC}{V}\right)\left(\frac{V - mC}{V - C}\right) \tag{8.15}$$

The probability that p minutiae among the n input template minutiae can be falsely matched with m minutiae in the target templates can be written as follows:

$$P_{3D}(m, n, p) = \binom{n}{p}\left(\frac{mC}{V}\right)\left(\frac{(m-1)C}{V-C}\right)\ldots\ldots\left(\frac{(m-p-1)C}{V-(p-1)C}\right)$$

$$\times \left(\frac{V-mC}{V-pV}\right)\left(\frac{V-(m-1)C}{V-(p+1)C}\right)\ldots\ldots\left(\frac{(V-(m-(n-p+1))C}{V-(n-1)C}\right) \tag{8.16}$$

The first line term in above equation represents the probability that p minutiae from input templates are matched, while the second line term represents the probability that the rest $(n–p)$ of the minutiae are not matched. We can rewrite (8.16) to compute the probability, when there can be exactly p minutiae that are matched, among n minutiae in template Q and m minutiae in template P, as follows:

$$P_{3D}(J, m, n, p) = \frac{\binom{m}{p}\binom{J - m}{n - p}}{\binom{J}{n}} \tag{8.17}$$

As stated earlier, we used (8.12–8.17) to estimate the false random correspondence probability between two 3D fingerprint templates while considering *only* the matching in spatial locations. When such p minutiae's positions are matched as in (8.17), the probability of q minutiae's direction θ being matched can be estimated as follows:

$$\binom{p}{q}(l)^q(1 - l)^{p-q} \tag{8.18}$$

where l is the probability of matching a 3D minutiae along orientation θ once they are falsely matched at 3D location (x, y, z). This probability or l is the same as computed in Eq. (8.10). The probability of matching q minutiae in both position (x, y, z) and direction θ can be written as follows:

$$P_{3D}(J, m, n, q) = \sum_{p=q}^{\min(m,n)} \left(\frac{\binom{m}{p}\binom{J-m}{n-p}}{\binom{J}{n}} \times \binom{p}{q}(l)^q(1-l)^{p-q} \right) \quad (8.19)$$

With q minutiae's position and direction θ already being matched, the probability of r minutiae's direction ϕ being matched can be estimated as follows:

$$\binom{q}{r}(k)^r(1-k)^{q-r} \quad (8.20)$$

where k denotes the probability of matching a 3D minutiae along orientation ϕ once they are falsely matched at 3D location (x, y, z) and at orientation θ. The k is the same as estimated from Eq. (8.11). The probability of matching r minutiae in two 3D fingerprint templates in both position (x, y, z) and spatial orientations θ and ϕ can be written as follows:

$$P_{3D}(J, m, n, r)$$

$$= \sum_{p=r}^{\min(m,n)} \sum_{q=r}^{p} \left(\frac{\binom{m}{p}\binom{J-m}{n-p}}{\binom{J}{n}} \times \binom{p}{q}(l)^q(1-l)^{p-q} \times \binom{q}{r}(k)^r(1-k)^{q-r} \right) \quad (8.21)$$

Above equation is the probability of false random correspondence, or the expected expression for the individuality of 3D fingerprints, where m and n denote number of 3D minutiae in two 3D fingerprints being matched, J represents the nearest integer corresponding to $J = V/(\frac{4}{3}\pi r_0^3)$, and $r_0 = \Delta r$ (Eq. 15 in Chap. 6), while l and k, respectively, denote the probability of matching a 3D minutiae along orientation θ and along orientation ϕ, respectively, *once* they are falsely matched at 3D location (x, y, z). Above estimation of false correspondence assumes that the angles ϕ and θ are independent of their location (x, y, z).

8.1.1 Relative Improvement from 3D Fingerprint Individuality

Assuming that the images from the same fingers are acquired from conventional 2D fingerprint sensors, we can estimate the expected improvement in the individuality of fingerprints using 3D fingerprint systems. The probability of false random correspondence using conventional minutiae representation in 2D space [7], i.e. matching q minutiae both in spatial location (x, y) and also along the direction θ, is given by

$$P_{2D}(M, m, n, q) = \sum_{p=q}^{\min(m,n)} \left(\frac{\binom{m}{p}\binom{M-m}{n-p}}{\binom{M}{n}} \times \binom{p}{q}(l)^q(1-l)^{p-q} \right) \quad (8.22)$$

where M represents the nearest integer for $M = A/(\pi r_0^2)$. Here, we assume that the *same* two source fingerprints are matched in 2D space and 3D space, and $J = M$, using (8.22) we can rewrite (8.21) as

$$P_{3D}(M, m, n, r) = \sum_{q=r}^{p} \left(\binom{q}{r}(k)^r(1-k)^{q-r} \right) \times P_{2D}(M, m, n, q) \quad (8.23)$$

Above equation suggests that the probability of false random correspondence using the 3D minutiae matching is expected to be smaller, as compared to the case when only *respective* 2D minutiae matching is employed. The extent of such reduction depends on k [0–1], i.e. nonlinearly decrease with the reduction in tolerance while matching angle ϕ. In practice, $J > M$ (which would further reduce the probability of false correspondence). It should, however, be noted that this assumption is only to compute upper bound on the probability of false correspondence for the comparison.

In order to quantitatively ascertain the uniqueness of 3D fingerprints, we can compute the probability (using 21) that any two arbitrary 3D fingerprints can be *falsely* matched when it is known that these two 3D fingerprints have been acquired from two different sources/fingers. We can use the 3D fingerprint matching criterion, i.e. tolerances, developed in Chap. 6 for illustrating the performance. Table 8.1 illustrates the probability of such false random correspondence for matching two 3D fingerprint minutiae templates, which uses same individuality model parameters empirically estimated for a 500 dpi sensor in [7], i.e. $l = 0.267$ with $k = 0.222$. The decision thresholds for Δr, Δ_{A_θ}, Δ_{A_ϕ}, for two 3D minutiae to be considered as matched, was estimated from our experiments. These threshold limits for Δr, Δ_{A_θ}, Δ_{A_ϕ} were 18, 20°, and 20° respectively. The new or the extreme right column in Table 8.1 illus-

Table 8.1 Comparative individuality of 2D and 3D fingerprints using correspondence probabilities

M	m	n	q	l	k	P(False random fingerprint correspondence)	
						2D [7]	3D
104	26	26	26	0.267	0.222	5.27e−40	5.51e−57
104	26	26	12	0.267	0.222	3.87e−9	8.94e−17
176	36	36	36	0.267	0.222	5.47e−59	1.68e−82
176	36	36	12	0.267	0.222	6.10e−8	1.90e−15
248	46	46	46	0.267	0.222	1.33e−77	1.21e−107
248	46	46	12	0.267	0.222	5.86e−7	2.46e−14
70	12	12	12	0.267	0.222	1.22e−20	1.77e−28

trates the possible improvement in the uniqueness of fingerprints when the same minutiae features are matched in 3D space. The results in this table indicate significantly enhanced limits in the number of persons that can be accurately identified using the 3D fingerprints over the currently believed such limits with the usage of conventional 2D fingerprint imaging.

8.2   Probability of False Random Correspondence Using Noisy Minutiae Matching

The key purpose of estimating individuality of 3D fingerprint matching in the previous section was to provide theoretical estimate on the performance of 3D fingerprint systems. Such estimate can serve as 'upper bound' on the performance, instead of *lower bound* or actual performance due to the effect of noise. The model introduced in the previous section is inherently simple and does not consider dependencies among the 3D features. This 3D fingerprint individuality model also does not account for errors in the localization and detection of 3D minutiae. Therefore, the individuality model for 3D fingerprints introduced in the previous section can be further improved by accounting for the observed variability in the 3D minutiae locations and directions, and is discussed in the following.

The probability of a random correspondence (PRC) for w or more minutiae matches between two 3D fingerprint templates I and T that can be estimated as follows:

$$\text{PRC}(w|m, n) = \sum_{u \geq w} p^*(u; I, T) \tag{8.24}$$

where p^* is the probability of observing exactly u minutiae which matches among the m minutiae recovered in template I and n minutiae recovered in template T. Reference

[10, 11] provides a closed-form expression for p^* using Poisson probability mass function with mean $\lambda(I, T)$,

$$p^*(w; I, T) = \frac{e^{-\lambda(I,T)}\lambda(I, T)^w}{w!}$$ (8.25)

$$\lambda(I, T) = mn\, p(I, T)$$

where

$$p(I, T) = P\left(\sqrt{(x'_i - x_j)^2 + (y'_i - y_j)^2 + (z'_i - z_j)^2} \leq r_0 \text{ and } \min(|\theta'_i - \theta_j|, 360\right.$$
$$\left. - |\theta'_i - \theta_j|) \leq \theta_0 \text{ and } \min(|\phi'_i - \phi_j|) \leq \phi_0\right)$$ (8.26)

Equations (8.9–8.11) assume that the probability of the features for a given minutia has uniform distribution in the given space. Reference [10] suggests finite mixture models for the minutiae variability in the fingerprint which fit the cluster of features representing the minutiae; therefore, $p(I, T)$ can be defined as follows:

$$p(I, T) = \iiint\limits_{(s_2,\theta_2,\phi_2)\in B(s_1,\theta_1,\phi_1)} f_I(s_1, \theta_1, \phi_1) f_T(s_2, \theta_2, \phi_2) ds_2 d\theta_2 d\phi_2 ds_1 d\theta_1 d\phi_1$$

(8.27)

where $s = (x, y, z)$ and B (s_1, θ_1, ϕ_1) is the space bounded by (r_0, θ_0, ϕ_0), while f is the G-component mixture distribution for the minutiae in 3D space.

$$f(s, \theta, \phi | \Theta_G) = \sum_{g=1}^{G} \tau_g f_g^X (s | \mu_g, \Sigma_g) \cdot f_g^D (\theta | v_g, \kappa_g, p_g) \cdot f_g^\phi (\phi | \eta_g, \zeta_g).$$ (8.28)

where τ_g is the weight and $f_g^X (s | u_g, \Sigma_g)$ is the density of Bivariate Gaussian random variable with mean μ_g and covariance matrix Σ_g. The second term can be estimated by a two-component mixture of von Mises distributions:

$$f_g^D (\theta | v_g, \kappa_g, p_g) = \begin{cases} p_g v(\theta), & \text{if } \theta \in [0, \pi) \\ (1 - p_g) v(\theta - \pi), & \text{if } \theta \in [\pi, 2\pi) \end{cases}$$ (8.29)

where $v(\theta)$ is the von Mises distribution

$$v(\theta) \equiv v(\theta | v_g, \kappa_g) = \frac{2}{I_0(\kappa_g)} e^{(\kappa_g \cos 2(\theta - v_g))}$$ (8.30)

and $I_0(\kappa_g)$ is

$$I_0\left(\kappa_g\right) = \int\limits_0^{2\pi} e^{\left(\kappa_g \cos(\theta - v_g)\right)} d\theta \tag{8.31}$$

where v_g and κ_g are the mean angle and the precision of von Mises distribution, p_g and $(1 - p_g)$ are the probability of 3D minutiae direction for θ and $\theta + \pi$. The last term in (8.28) is the distribution function for ϕ which is defined by another von Mises distribution with mean angle η_g and precision ζ_g. When low-quality 3D fingerprint images are used for matching, the PRC is expected to be higher. Therefore, similar to as in [10], we can model the relationship of $\mathrm{PRC}(w|m, n)$ with the image quality as follows:

$$\log p(I, T) = \mu + \gamma_{(q(I),q(T))} + \beta_{ch(I)} + \beta_{ch(T)} + \alpha_{\mathrm{Finger}(I)} + \alpha_{\mathrm{Finger}(T)} + \varepsilon(I, T) \tag{8.32}$$

where μ is the overall mean of the logarithm of $p(I,T)$, $\gamma_{(q(I),q(T))}$ is the effect on the image quality on template I and T, $q(S)$ is the quality measure for 3D fingerprint, $ch(S)$ represents the characteristic of fingerprint class, $\varepsilon(I, T)$ is the distribution of random error with zero mean and variance σ^2. The Finger(S) represents variation in 3D reconstruction for impression S, from the same finger (I and I'), which is expected to have some correlation as follows:

$$\mathrm{Corr}\left(\log p(I, T), \log p\left(I', T'\right)\right) = \frac{\sigma_\alpha^2}{2\sigma_\alpha^2 + \sigma^2} = \frac{2}{2 + v} \tag{8.33}$$

for $v = \sigma^2/\sigma_\alpha^2 > 0$. The individuality of 3D fingerprint under noisy 3D minutiae can be represented using (8.24) and describes how PRCs can be influenced with the degradations in the quality of 3D fingerprints.

References

1. Stony D (2001) Measurement of fingerprint individuality. In: Lee HC, Gaensslen RE (eds) *Advances in fingerprint technology*. CRC Press
2. Srihari S, Srinivasan H Individuality of fingerprints: comparison models and measurements. CEDAR Technical Report TR-02-07. http://www.cedar.buffalo.edu/~srihari/papers/TR-02-07. pdf
3. Osterburg J (1997) Development of a mathematical formula for the calculation of fingerprint probabilities based on individual characteristics. J Am Statistical Ass 772:72
4. Roxburgh T (1933) On the evidential value of fingerprints. Sankhya: The Indian J. Statistics 1:89
5. Henry E (1900) Classification and uses of fingerprints. Routledge & Sons London, pp 54
6. Champod C, Margot P (1996) Computer assisted analysis of minutiae occurrences on fingerprints. In: Almog J, Spinger E (eds) Proceeding international symposium fingerprint detection and identification. pp 305
7. Pankanti S, Prabhakar S, Jain AK (2002) On the individuality of fingerprints. IEEE Trans Pattern Anal Mach Intell 24

8. Kücken M, Newell AC (2005) Fingerprint formation. J Theor Biol 235:71–83
9. Hale A (1952) Morphogenesis of volar skin in the human fetus. Am J Anatomy 91:147–173
10. Dass SC (2010) Assessing fingerprint individuality in presence of noisy minutiae. IEEE Trans Info Forensics & Security 5(1):62–70
11. Zhu Y, Dass SC, Jain AK (2007) Statistical models for assessing the individuality of fingerprints. IEEE Trans Info Forensics & Security 2(3):391–401

Index

© Springer Nature Switzerland AG 2018   121
A. Kumar, *Contactless 3D Fingerprint Identification*, Advances in Computer Vision
and Pattern Recognition, https://doi.org/10.1007/978-3-319-67681-4

Printed in the United States
By Bookmasters